Springer Tracts in Electrical and Electronics Engineering

Springer Tracts in Electrical and Electronics Engineering (STEEE) publishes the latest developments in Electrical and Electronics Engineering - quickly, informally and with high quality. The intent is to cover all the main branches of electrical and electronics engineering, both theoretical and applied, including:

- Signal, Speech and Image Processing
- Speech and Audio Processing
- Image Processing
- Human-Machine Interfaces
- Digital and Analog Signal Processing
- Microwaves, RF Engineering and Optical Communications
- Electronics and Microelectronics, Instrumentation
- Electronic Circuits and Systems
- Embedded Systems
- Electronics Design and Verification
- Cyber-Physical Systems
- Electrical Power Engineering
- Power Electronics
- Photovoltaics
- Energy Grids and Networks
- Electrical Machines
- Control, Robotics, Automation
- Robotic Engineering
- Mechatronics
- Control and Systems Theory
- Automation
- Communications Engineering, Networks
- Wireless and Mobile Communication
- Internet of Things
- Computer Networks

Within the scope of the series are monographs, professional books or graduate textbooks, edited volumes as well as outstanding PhD theses and books purposely devoted to support education in electrical and electronics engineering at graduate and post-graduate levels.

Review Process

The proposal for each volume is reviewed by the main editor and/or the advisory board. The books of this series are reviewed in a single blind peer review process.

Ethics Statement for this series can be found in the Springer standard guidelines here https://www.springer.com/us/authors-editors/journal-author/journal-author-helpdesk/before-you-start/before-you-start/1330#c14214

Abhishek Srivastava · Parasharam M. Shirage

Perovskite Solar Cells

Fundamental Approaches and Materials Processing for Sustainable Future

Springer

Abhishek Srivastava
Department of Metallurgical Engineering
and Materials Science (MEMS)
Indian Institute of Technology Indore
Indore, Madhya Pradesh, India

Parasharam M. Shirage
Department of Metallurgical Engineering
and Materials Science (MEMS)
Indian Institute of Technology Indore
Indore, Madhya Pradesh, India

ISSN 2731-4200 ISSN 2731-4219 (electronic)
Springer Tracts in Electrical and Electronics Engineering
ISBN 978-981-97-7623-8 ISBN 978-981-97-7624-5 (eBook)
https://doi.org/10.1007/978-981-97-7624-5

This Springer imprint is published by the registered company Springer Nature Singapore Pte Ltd.
The registered company address is: 152 Beach Road, #21-01/04 Gateway East, Singapore 189721, Singapore

Preface

Welcome to the fascinating world of next-generation *perovskite solar cells* and their exciting applications! In this book, we embark on a remarkable journey that explores the cutting-edge advancements and potential of perovskite solar technology, poised to revolutionize the field of renewable energy.

Perovskite solar cells have emerged as a game-changer in the quest for efficient and sustainable solar power. These remarkable devices are constructed using a unique class of materials with a crystal structure that mimics the naturally occurring mineral known as perovskite. Harnessing the extraordinary photovoltaic properties of perovskite, researchers and scientists worldwide have unleashed a wave of innovation that promises to transform our energy landscape.

As you dive into the pages of this book, you will uncover the underlying principles and mechanisms that make perovskite solar cells so exceptional. From their fascinating optical and electronic properties to their remarkable versatility, we will delve deep into the science behind these devices. We will explore the methods and techniques used to fabricate perovskite solar cells, including the latest breakthroughs in materials synthesis, device architectures, and manufacturing processes. But the story doesn't end there. The true power of perovskite solar cells lies in their vast range of applications. Throughout this book, we will explore the diverse fields where perovskite solar cells are poised to make a significant impact. From building-integrated photovoltaics and wearable electronics to automotive integration and space exploration, we will unravel the immense potential of perovskite solar technology across various industries.

Moreover, we will examine the challenges and obstacles that researchers face as they strive to optimize the stability, durability, and scalability of perovskite solar cells. This book serves as a comprehensive guide, offering insights into the latest strategies and innovations that aim to overcome these hurdles, paving the way for commercialization and widespread adoption. We have assembled a team of esteemed experts and visionaries who have dedicated their careers to perovskite solar cells and their applications. Their invaluable contributions, coupled with the tireless efforts of countless scientists around the world, have propelled this field forward, propelling it into the forefront of renewable energy research.

Whether you are an aspiring researcher, a seasoned scientist, a curious student, or an industry professional, this book is designed to provide you with a comprehensive understanding of next-generation perovskite solar cells. It is a testament to the remarkable progress made and the boundless potential that lies ahead.

So, embark on this exciting journey with us as we explore the world of next-generation perovskite solar cells and their ground-breaking applications. Together, let us envision a future powered by clean, abundant, and sustainable energy.

Mr. Abhishek Srivastava
Senior Research Fellow (DST INSPIRE)
Department of Metallurgical Engineering and Materials Science (MEMS)
Indian Institute of Technology Indore
Indore, Madhya Pradesh, India

Prof. Parasharam M. Shirage
RJNF, FJSPS, FMASc, MNASc
Department of Metallurgical Engineering and Materials Science (MEMS)
Indian Institute of Technology Indore
Indore, Madhya Pradesh, India
pmshirage@iiti.ac.in

Acknowledgements The authors are delighted to thank the Indian Institute of Technology Indore for providing a very supportive and excellent research environment. The authors wish to thank DST, SERB, AICTE, MHRD, and other government agencies for providing research funds to conduct the research work from time to time.

Contents

About the Authors

Mr. Abhishek Srivastava is presently pursuing his Ph.D. in the Department of Metallurgical Engineering and Materials Science at IIT Indore, India. His prolific body of work spans over 18 peer-reviewed publications, covering advancements in third-generation solar cells, including dye-sensitized solar cells (DSSCs), perovskite solar cells, and novel photovoltaic materials. Additionally, he optimizes the optoelectronic properties of solar radiation-absorbing materials, aiding photovoltaic efficiency. His interests extend to stable stoichiometries for efficient photovoltaic applications and sustainable technologies such as green hydrogen generation and wastewater remediation. He has published numerous articles cited by peers in the scientific community, reflecting his impact on sustainable energy and advanced materials research with an h-index of 11 and i10-index of 11. Recently, he received the "Outstanding Paper Award-2023" from the esteemed Energy Advances journal published by the Royal Society of Chemistry. With a focus on sustainable, environmentally friendly technologies, he has significantly contributed to material science, particularly in developing non-toxic, efficient alternatives for solar energy applications. His expertise in surface passivation, hybrid nanofluids, and photoelectrochemical systems highlights his innovative approach to improving the efficiency and durability of solar technologies. Mr. Abhishek's ongoing research includes the integration of flexible solar cells and Building Integrated Photovoltaics (BIPVs), which aim to drive the next wave of sustainable energy solutions.

Prof. Parasharam M. Shirage is a full Professor in the Department of Metallurgical Engineering and Materials Science (MEMS) at the Indian Institute of Technology Indore and a Visiting Professor at Meiji University, Ikuta Campus, Tokyo, Japan. He is currently serving as Vice-Chairman of JEE (Advanced) at IIT Indore. He has been listed in the world's top 2 % of scientists since its inception and is a gold medallist in materials science at IIT Indore. He has more than 13 years of teaching experience at IIT Indore. He has served as Dean of Planning, Head of the Department of Metallurgical Engineering and Materials Science, and Chief Warden of the Hall of Residence at IIT Indore from 2017 to July 2019. Prior to that, he was a Visiting Fellow at the Tata Institute of Fundamental Research (TIFR), Mumbai; he was a Senior Scientist at the National Institute of Advanced Industrial Science and Technology (AIST), Tsukuba, Japan. He authorizes over 450 peer-reviewed research papers with 7499 citations h-index 50, i10-index = 164. Under his supervision, 19 Ph.D., 41 master's, and 15 B. Tech students have completed their thesis. Currently, his research group has 6 B. Tech, 7 Ph.D. students, and 1 master's student. He is the recipient of a highly prestigious Medal from the Materials Research Society of India (MRSI) 2020. He is a Fellow of the Maharashtra Academy of Sciences and a member of the Indian Association of Physics Teachers (IAPT). Among the various honours and awards received, mention must be made of the MRSI Medal 2020, Excellence in Teaching i.e., Best Teacher Award at IIT Indore (2016), Japan Society for the Promotion of Science (JSPS) award from Japan (2006), MEXT fellowship (2008) from Japan, MK-21 (Geongnam Technopark) award from Ministry of South Korea (2004), Visiting Fellowship from TIFR Mumbai, Best paper presentation award from Materials Research Society of India (2000 and 2019), RSC outstanding paper 2024, Paper of Editors suggestion from American Physical Society three times (2011, 2009, 2008), Paper of editor's choice from Physical Society of Japan (JPS) (2011, 2009). He is a fellow of the Institute of Physics, U.K., a member of the executing body of the Indian Scientist Association in Japan (ISAJ), on the editorial board of The Open Applied Physics Journal (BENTHAM), and a reviewer for APS, ACS, JPS, IOP, Elsevier journals.

List of Figures

List of Tables

Chapter 1
Perovskite Solar Cells: Futuristic Reliable Renewable Energy Technology

Abstract This chapter provides a historical overview of photovoltaic technology, charting its evolution from intermittent viability to a crucial solution for contemporary energy needs. Emphasizing its cost-effectiveness, the analysis explores diverse applications, from space technology to public illumination and large-scale photovoltaic parks. It provides typical insights into the solar cell operational principles, such as p–n junctions. It reviews various models, including 1M3P, 1M5P, and complex structures like 1M7P, where M and P stand for model and parameters, respectively. The study uses electrical models and a thorough literature assessment to provide insights into the performance variations of solar cells under challenging conditions such as variation in temperature, irradiation changes, degradation, and shading, offering new research endeavors to address real-world problems. This chapter serves as a concise overview, aiming to engage new researchers by highlighting historical milestones, foundational principles, and notable applications in the dynamic field of photovoltaics.

Keywords Next-generation photovoltaics · Si solar cells · Thin-film solar cells · Hybrid solar cells · Perovskite solar cells · Degradation mechanism · Electrical modeling: 1M3P · 1M5P and 1M7P

1.1 Introduction

In combatting climate change, it is crucial to have the ability to produce electricity from renewable energy sources. As the sun emits 4 million exajoules of solar radiation onto the earth's surface every year, solar radiation is often viewed as the most ubiquitous and reliable energy source (Kabir et al. 2018). In fact, within an hour of sunlight radiation on the earth's surface, it can produce more energy than consumed in a year. Over the past ten years, the demand for clean energy has skyrocketed, especially in solar photovoltaics (PV). Solar PV deployment is increasing yearly, as in 2018, more than 100 GW of solar PVs were installed globally. According to the International Energy Agency (IEA), this yearly installation capacity will increase

A. Srivastava and P. M. Shirage, *Perovskite Solar Cells*, Springer Tracts in Electrical and Electronics Engineering, https://doi.org/10.1007/978-981-97-7624-5_1

to 3000–4000 GW solar PVs by 2030. The solar PV capacity is reaching 709 GW, nearly 24.3% of global capacity (IRENA 2022). The success of solar PVs was due to their longer lifespan, practical applications, and meager maintenance cost with super performance. Unlike other renewable energy sources, solar PV can be deployed and used in urban domains such as building-integrated photovoltaic (BIPV) and building-applied photovoltaic (BAPV). To maximize the amount of energy that can be extracted from solar irradiation, a significant scientific effort has been taken to increase the performance of solar panels (Reddy et al. 2020; Zhang et al. 2018; Ghosh et al. 2020).

The PV technology has captured the interest of scientists worldwide since it is the most effective renewable energy capacity to cater to future energy demands. At the moment, silicon PV technology rules the market. The fact that silicon PV technology alone will not be able to supply the world's energy needs in the future has forced researchers to look for more effective PV alternatives, which has drawn their attention to PSCs. According to the photovoltaic effect, this limitless sunlight irradiance can be transformed into electricity using a semiconductor material (Srivastava et al. 2023b). As the photovoltaic operating theory suggested, an electron is transported from the valence band to the conduction band by electromagnetic radiation from the sun, and an electron–hole pair is formed. Due to the formation of electron–hole pairs, an electrical current can be produced if those excited electrons are supplied back to the valence band through an external circuit (Shockley and Queisser 2018). Besides all the necessities, a maximum limit is defined for the power conversion efficiency (PCE) of a single-junction solar cell, known as the Shockley-Queisser limit. According to this theory, the maximum 33.16% PCE was obtained with the solar cell having an active material of 1.34 eV bandgap under the standard illumination of AM 1.5G (Rühle et al. 2016). However, extensive research is going on to maximize the power conversion ability of the functional materials.

1.2 History of Photovoltaics

In 1839, the French physicist Alexandre Edmond Becquerel observed the photovoltaic effect while conducting electrochemical experiments. During his experimentation, he noted the phenomenon of electricity generation in silver and platinum electrodes exposed to sunlight. This observation marked the discovery of the photovoltaic effect, a process in which light energy is converted into electrical energy through the interaction of photons with semiconducting materials (Andreev et al. 2003).

In 1873, Willoughby Smith communicated the manifestation of the photovoltaic effect in selenium to his colleague Latimer Clark. Smith's experimentation with selenium was motivated by the need for high-resistance materials for testing and signaling in the submersion of long submarine cables. He hermetically sealed selenium bars in glass tubes with platinum wires for connection and observed significant changes in their resistance with varying light intensity. After four years, in 1877, Adams and Richard Day created the first solar cell with a PCE of about 0.5% using selenium. Subsequently, Charles Fritts increased the device's efficiency by employing a selenium wafer between two thin metal layers (Carmo et al. 2021).

In 1905, Albert Einstein proposed the photoelectric theory, which incorporated concepts previously introduced by Max Planck. This theory established a relationship between light waves and photons, the discrete packets of energy that constitute light, and described the linear relationship between photon energy and frequency, with Planck's constant serving as the proportionality constant. Einstein received the Nobel Prize in Physics in 1921 due to his revolutionary research, which included the development of this theory. His contributions to theoretical physics, particularly the discovery of the law of the photoelectric effect, which clarified the minimal frequency at which photons must release electrons, were acknowledged by the Nobel Prize Academy (Einstine 1920).

The replacement of selenium with silicon led to a significant improvement in the efficiency of solar cells. Russell Ohl identified the silicon's n- and p-type regions in 1939, opening the door for the photoelectric effect in p–n junctions. The development of other important technologies, such as the transistor and the solar cell, is the subject of this review article, which was made possible by this achievement. In 1940, Ohl created the first silicon solar cell, a light-sensitive electric device, at Bell Labs (Ohl 1946). Researchers and innovators noticed in early trials that the current produced by these devices was proportionate to the incident radiation, with the wavelength of the radiation determining the proportionality. These devices were utilized as sensors instead of solar cells due to their low efficiency. Instead of using solar cells, the German engineer Werner Siemens, the company's founder, was the first to market these innovative devices as light sensors for photography.

It wasn't until 1954 that chemist Calvin Fuller of Bell Labs developed a process for doping silicon, which involved intentionally introducing impurities into the material. This breakthrough enabled the fabrication of the first known p–n junction diode, developed by Fuller's colleague and physicist Gerald Pearson. Pearson created the hub by immersing a doped silicon bar in lithium, which resulted in the observation of photovoltaic properties. Using boron and arsenic dopants led to an impressive efficiency of 6% (Nokia Bell Labs 1954).

The first use of a solar cell was in Americus, Georgia, in 1955 when one was used as the power supply for a telecommunications network. This application came about after the gadget was shown on April 25, 1954, in Washington, D.C., at the USA National Academy of Sciences annual meeting. Other applications of such a solar cell have emerged at an extraordinary pace. NASA launched the Vanguard-I satellite, the first to use a solar power system, on March 17, 1958. The satellite had two transmitters: a backup that used a solar system made up of 6-silicon solar cells and a primary transmitter that used a mercury cell source. With a lifespan of only 20 years for the typical mercury power system, the 8 years that the solar cells operated was an amazing feat. NASA was first dubious about the benefits and eventual need for a solar power system on their extraterrestrial products. Nevertheless, photovoltaic technology demonstrated its value as this novel approach to power generation enabled the success of multiple missions. The Russian space program launched Sputnik-3, a satellite with a photovoltaic solar source, two months after Vanguard-I. Photovoltaic power sources have been essential for space endeavors ever since (Bernardes et al. 2021).

The space race was a significant factor in advancing photovoltaic technology. However, the high cost of commercialization and the need for optimized manufacturing processes posed substantial barriers to the technology's widespread public application. Former Communications Satellite Corporation (Comsat) employee Joseph Lindmeyer created a method in the 1970s that raised silicon solar cells' efficiency by 50%. In order to create solar cells for use in the general public, Lindmeyer and Peter Varadi formed "Solarex" in 1973. Despite the patent owned by Comsat instead of Lindmeyer or Solarex, the company gained approximately 50% of the photovoltaic industry market by 1980 due to the rising interest in photovoltaic technology following the oil crisis. After 1973, newer technologies—like poly-Si (polycrystalline silicon) and a-Si (amorphous silicon)—that employed less expensive and labor-intensive techniques took the place of mono-Si (monocrystalline silicon).

Since the discovery of photovoltaic technology, various advancements have been made using different materials and techniques, leading to the categorization of solar cells into two and, later, three generations. Recently, there has been a proposal for a fourth categorization. Advancements in other fields, such as electronics, photonics, and quantum mechanics, have also contributed to improving photovoltaic cells. For instance, flexible and painted solar cells have been developed. From 1976 till the present, the National Renewable Energy Laboratory (NREL) has kept track of the best-verified PCEs for research cells across all photovoltaic technologies (NREL 2023).

Including other elements in the photovoltaic system, such as inverters, batteries, and cables, has also contributed to its overall efficiency and effectiveness. Notably, advancements in any of these components have improved the system's efficiency, increasing the benefits of photovoltaic technology.

1.2.1 Classification of Photovoltaics

Generally, solar cells are classified into three generations based on the materials and production techniques utilized during fabrication, as shown in Fig. 1.1. The first generation, the most widely used in photovoltaic modules today, is wafer-based and relies on established techniques used in integrated circuit manufacturing, particularly in silicon wafer production. This familiarity with existing manufacturing processes has allowed for significant efficiencies and cost savings in production. Second-generation solar cells were developed to reduce costs and enhance the first-generation solar cells performance.

The industry that produces solar modules strives to keep the ratio of the module's total cost to that of its installation half. On the other hand, the cost of the cells ought to be less than half that of the module. Reducing the high material demand of the cells and the associated costs was necessary to achieve a cheaper module. As

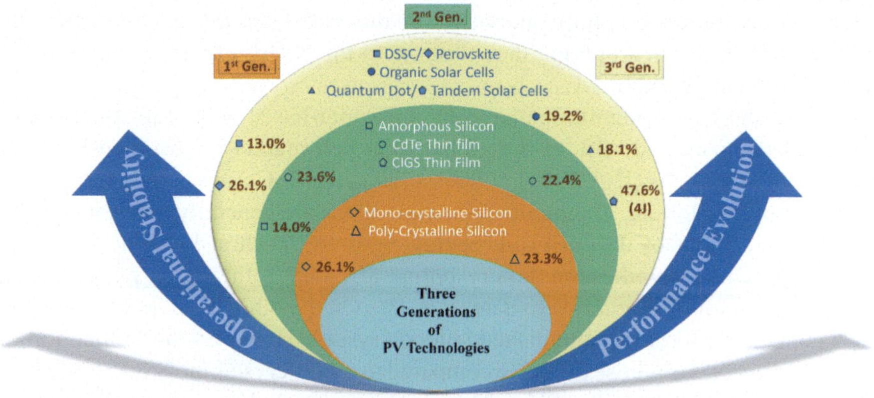

Fig. 1.1 Schematic illustration of all three generations of photovoltaic technologies

a result, the second generation of cells was developed using thin-film technology, characterized by a thin absorber layer of a few micrometers, compared to the 100–200 μm of the first generation. Nevertheless, efficiency suffered as a result of this compromise. With a thin substrate, the efficiency levels of the first generation were initially unattainable. However, other manufacturing techniques, like spray coating onto a thin glass substrate and other vacuum-based technologies, which needed lower production temperatures, were used to offset the drop in efficiency (da Silva Ravasco et al. 2021). Additionally, flexible, thin-film cells open up new possibilities for use, such as wearable electronics. They are also a greener option because of their shorter energy payback period and lower greenhouse gas emissions. The early efficiency disadvantage has vanished, with records exceeding 20% and a current average of 23.4%.

Third-generation solar cells are not like those of the previous generation in that they do not depend on a typical p–n junction. Rather, they are constructed using novel and distinct materials like organic compounds and thus the name. The generation's capacity to adjust bandgap energies in response to compositional modifications is another noteworthy advancement and a necessary component of multi-junction cell production. Notably, the realized efficiency levels and areas of research focus for each type of solar cell might vary significantly, even while technologies from the same generation may have some properties. As a result, a metric to evaluate investment and research interest can be developed. A Research Tendency (RT) is identified by examining the efficiency breakthroughs for every type of cell from the National Renewable Energy Laboratory (NREL) chart and contrasting it with the relevant time period. Efficiency records are usually associated with advances in fabrication methods, which can only occur when there is enough research done on the subject. But RT by itself is not enough to evaluate current research interest because technology may have made multiple strides forward only to peter out in its advancements. As a result, RT_5, a different complimentary statistic that only takes the last five years into account, is introduced. Applying these criteria to the NREL efficiency chart yields the results shown in Table 1.1.

Most of the first-generation solar cells are wafer-based cells with thicknesses of several 100 μm, and 90% of the global solar cell market is governed by these wafer-based solar cells (IEA 2010). Silicon solar cells (SSCs) are mainly monocrystalline or polycrystalline. The mono-crystalline SSCs are generally fabricated using single-crystal silicon by the costly and energy-rigorous Czochralski method (Sharma et al. 2015). Currently, these mono-crystalline SSCs are saturated in performance, and the maximum PCE attained to date is 26.6%. On the other hand, polycrystalline SSCs contain numerous distinguished silicon crystals. However, the fabrication of polycrystalline SSCs is cost-effective compared to mono-crystalline SSCs; its performance is comparably less. The PV industry is still dominated by silicon cells made of monocrystalline and polycrystalline wafers. However, the cost of producing ultrapure semiconductor wafers is high in energy and cost (Yoshikawa et al. 2017).

Amorphous and polycrystalline semiconducting material is placed over the substrate to form the foundation of the second generation of cells. Especially in comparison with the first generation of cells, such solar cells are inexpensive per

Table 1.1 Advances and recent research trends of various solar cell technologies

Cell	Type	RT	RT_5	η (%)
Multi-junction solar cells	3J	1.385	0	44.4
	2J	0.211	0.4	35.5
	$4J^+$	0.750	0.6	47.1
GaAs	Single crystal	0.182	0.4	27.8
	Concentrator	0.182	0.4	30.5
	Thin-film crystal	0.533	0.4	29.1
Silicon	Single-crystal	0.088	0	27.6
	Multi-crystalline silicon	0.405	1.2	23.3
	Silicon heterostructures	0.550	0.6	26.7
	Thin-film crystals	0.450	0	21.2
Thin film	CIGS	0.150	0	23.3
	CdTe	0.533	0	22.1
Emerging	DSSCs	0.300	0.4	13.0
	Perovskite	1.500	1.6	25.8
	Perovskite/Si tandem	1.500	1.2	29.5
	Organic	1.200	1.8	18.2
	Inorganic (CZTSSe)	0.500	0.2	13.0
	Quantum dot	1.182	1.2	18.1
	Perovskite/CIGS tandem	1.000	0.6	24.2

Reproduced from Ref. (NREL.), Copyright (2023) with permission of NREL unit of energy supplied. The market acceptability of the cell has been impeded by numerous concerns related to the environment, mass production, materials, and technical aspects (Chopra et al. 2004). Thin-film solar cells, often referred to as second-generation solar cells, encompass materials like gallium arsenide (GaAs), copper indium gallium di-selenide (CIGS), and cadmium telluride (CdTe). Since minimum material is required, thin-film cells are more economically feasible. As these cells typically have a thickness of 1 μm compared to silicon wafers (350 μm), they can also be applied to curved surfaces.

GaAs achieved the maximum efficiency in this category of materials, with a PCE of 29.1%. CdTe and CIGS, which achieve 22.1% and 23.35% efficiencies, respectively, are two other important inclusions (Green et al. 2021; Nakamura et al. 2019). However, each of the aforementioned materials demonstrates drawbacks that would prevent widespread marketing. Due to their pricey and challenging production process, GaAs is only efficient in high-performance applications. Because they use hazardous substances like cadmium and rare elements like indium and tellurium, CIGS and CdTe are constrained by the materials that must be used. Developing thin-film technology to achieve great efficiency at an affordable expense is ongoing (Das et al. 2019). Hence, it can be concluded that Si technology has effectively dominated thin-film technology. The ability to roll and fold thin-film solar cells reduces the chance of falling, which is an advantage. They are small, lightweight cells that can produce electricity in low- and high-intensity solar irradiance. Wearable electronics and BIPVs utilize thin-film solar cells more frequently than PV plants.

While still under research, third-generation solar cells have shown promising possibilities and are advancing quickly. These materials are appealing because they can be produced using low-energy procedures and inexpensive materials. Third-generation solar cells of several kinds, including dye-sensitized, quantum dot, and perovskite solar cells (PSCs), are being studied. The PSCs are known as the most promising third-generation solar cells and has a commercial future. The PCE for PSCs increased from 3.5 to 25.8% over the last ten years (Liu et al. 2022; Wang et al. 2021). Generally, emerging organic photovoltaic technology (OPVs), copper zinc tin sulfide (CZTS), quantum dot cells, dye-sensitized solar cells (DSSC), and PSCs are referred to as third-generation photovoltaics. In the upcoming section, all three solar cell generations are discussed in detail.

1.2.2 First-Generation Solar Cells

1.2.2.1 Silicon Solar Cells

Photovoltaic technology has a long history, with silicon solar cells being one of the earliest types of solar cells. Despite the development of alternative photovoltaic technologies, silicon-based cells remain the most commonly used and prevalent.

This is primarily due to two factors: the favorable bandgap energy of silicon, which is 1.17 eV, and the abundance of high-quality silicon material available due to the already scaled-up production of silicon-based semiconductors for microchip manufacturing (Gordillo et al. 1992).

In astrophysics, one of the primary methods for studying newly discovered stars is analyzing their emission spectra, as they provide valuable insights into the star's composition and temperature. A star's emission spectrum is a band spectrum, with each band denoting the elemental bandgap energy values. Furthermore, there is a relationship between the temperature of the star and particle excitation: higher temperatures are associated with higher particle energy levels and excitation. A star with a greater surface temperature will produce a visible light spectrum that is blue in color due to its peak being moved toward the blue side. On the other hand, a cooler star produces a spectrum that has a peak that is moved toward the red end, giving the star a red color. The sun's emission spectrum peaks in the visible region toward green at a wavelength of 520 nm, with significant emissions in the yellow and red wavelengths as well.

In the design of photovoltaic devices, the bandgap energy of the semiconductor absorber is crucial in determining the efficiency of converting incident photons into charge carriers. The efficiency of this conversion is influenced by the sun's emission spectrum, which determines the optimal range for semiconductor bandgap energies that can yield high conversion efficiency. This optimal range is typically between 1.1 and 1.7 eV for homo-junction devices. Although silicon has a bandgap energy of 1.17 eV, which falls within this range, its efficiency is reduced due to its indirect bandgap nature. Auger recombination is the most common intrinsic recombination process in indirect bandgap semiconductors, which have a difference in momentum between the boundaries of the valence and conduction bands. This results in greater thermalization losses and energy dissipation during recombination. For effective solar energy conversion, direct bandgap semiconductors with larger bandgap values are therefore typically preferred. But how photovoltaic cells are developed was mostly determined by the microprocessor industry. Even though silicon's qualities may have been improved, the cost of silicon was more favorable than creating a specific semiconductor production process for solar cells because high-grade silicon wafers were already produced. Moreover, the photovoltaic industry could purchase lower-grade silicon wafers at a lower cost, which needed to be more suitable for integrated circuits. Consequently, more desirable semiconductors, like GaAs, were restricted to uses where certain characteristics were more crucial than price, like satellite power sources in orbit.

c-Si solar cells have seen a remarkable increase in efficiency of 20.1% since their inception, from an initial 6% to the current high of 26.1%. PCE has also been greatly impacted by the advances in semiconductor manufacture that have led to an increase in microprocessor computing capacity as bulk flaws in wafers have been gradually eliminated. But strong bulk quality alone is not enough for high-efficiency cells; over time, additional advancements in manufacturing have also helped to drive up efficiency. Wafer processing has seen a number of significant innovations, including multi-wire sawing, which lowers material loss and overall cell cost, and block

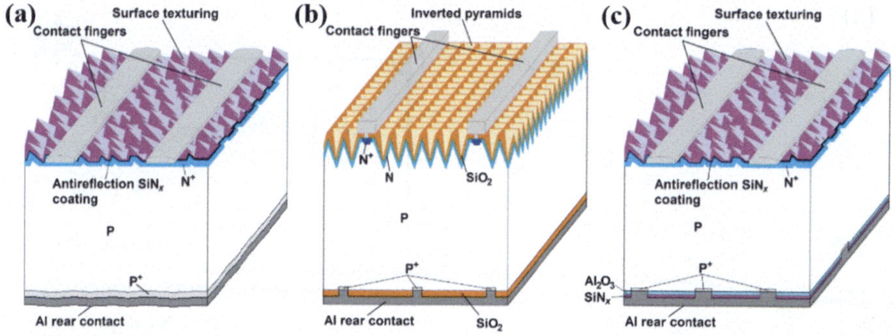

Fig. 1.2 Progresses made in developing Si solar cells. **a** Schematics of BSF-, **b** PERC-, and **c** modified PERC-structures of *c*-Si solar cells. Reproduced from Ref. (Benda et al. 2020), Copyright (2020) with permission of Elsevier

casting, which yields high-grade polycrystalline-silicon wafers. Still, at a cheaper cost, surface texturing reduces reflection from the top and bottom surfaces. The creation of a passivated emitter and rear cell (PERC) has further decreased the rear surface recombination velocity, which has also been effectively decreased by the addition of an aluminum back surface (Al-BSF). Schematic structures of these *c*-Si solar cells are depicted in Fig. 1.2 (Benda et al. 2020).

Currently, 95% of the world's solar energy production is produced by *c*-Si cells, demonstrating their supremacy in the industry. However, this figure is expected to shift with the recent advances in thin-film solar cell technology, which has demonstrated comparable efficiency to *c*-Si cells and the added benefit of reduced semiconductor costs. Furthermore, heterojunction-based complex architectures have outperformed *c*-Si homo-junction efficiency records. It is anticipated that these heterojunctions will become more significant in the future of photovoltaics, assuming a decrease in production costs, as technology and related procedures grow more efficient.

1.2.3 Second-Generation Solar Cells

1.2.3.1 Copper–Indium–Gallium–Selenide (CIGS)

Silicon (Si), the primary component of the photovoltaic market, has seen a significant drop in price in recent years. As a result, newer technologies need help in gaining a significant market share. Targeting the expensive parts of Si module manufacturing and creating solutions appropriate for various applications are essential if Si is to be effectively competed with. Achieving an ideal balance of system costs and reducing the significant dependence on semiconductor materials are two critical areas to improve in this regard. Schematic structures of these thin-film solar cells have been shown in Fig. 1.3a, b (Metzger et al. 2019; Kim et al. 2007).

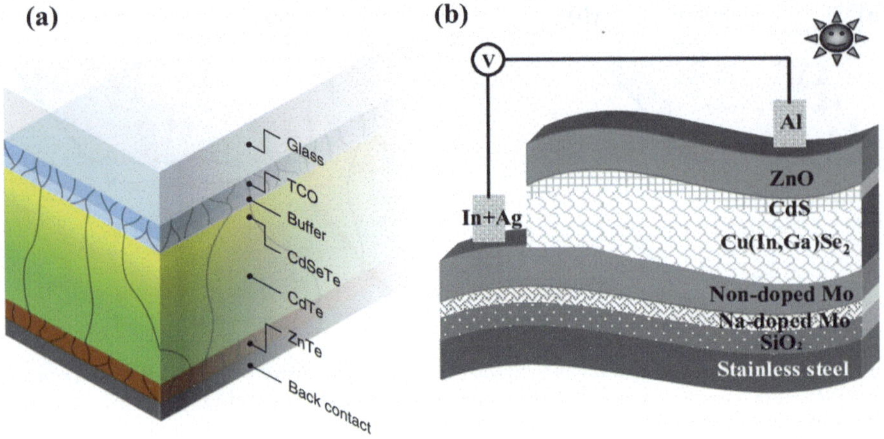

Fig. 1.3 Schematic structures of **a** CdTe, Reproduced from Ref. Metzger et al. (2019), Copyright (2019) with permission of Springer Nature. **b** Flexible CIGS thin-film solar cells, Reproduced from Ref. Kim et al. (2007), Copyright (1988) with permission of Trans Tech Publications

Second-generation photovoltaic technologies that use thin films, including copper indium gallium selenide (CIGS) have emerged to address these challenges. By using thin films, CIGS can reduce its dependency on semiconductor materials, which can be expensive and challenging to manufacture. This, in turn, can lower the overall cost of producing CIGS modules, making them more competitive with Si. Additionally, the flexibility of CIGS modules allows them to be utilized in various applications, such as building-integrated photovoltaics, where they can be integrated into building materials like roofs and facades. Overall, developing second-generation photovoltaic technologies, such as CIGS, is a critical step toward reducing the cost and increasing the versatility of solar power. This can ultimately enhance its viability as an alternative to conventional fossil fuel-based energy sources (Alves dos Santos et al. 2021).

Thin films of chalcopyrite materials, such as copper indium di-selenide ($CuInSe_2$) and its derivatives, have been investigated for their potential use in solar cells. Developing chalcopyrite thin films for solar cells builds upon earlier research on GaAs laser photodetectors. A $CdS/CuInSe_2$ heterojunction was studied for full spectrum quantum efficiency, which suggested the possibility of its application in solar cells. Modifications to this heterojunction resulted in an overall PCE of 12%. CIGS thin films were subsequently created by introducing larger bandgap elements, including CuInS, into the pre-existing thin films in order to raise the open-circuit voltage. Bandgap tailoring, which usually manifests as a grading in high-efficiency cells, was made possible by the integration of Ga by adjusting the Ga and In composition ratio in the range of 1.0 eV (in pure $CuInSe_2$) to 1.7 eV (in pure $CuGaSe_2$). These scientific developments demonstrate the potential of chalcopyrite thin films, particularly CIGS, for use in solar cells. By manipulating the composition of these materials, it is possible to optimize their properties for enhanced solar cell efficiency (Feurer et al. 2017).

A sequence of sputtering deposition steps is involved in the typical fabrication procedure for CIGS solar cells, which is then followed by mechanical or laser scribing. To create the back contact, a molybdenum layer is first sputtered onto a glass substrate. The CIGS absorber layer is then deposited by sputtering or co-evaporation, after which the precursors are sulfurized and seleniumized. This is followed by introducing a buffer layer, usually of cadmium sulfide, and then applying a transparent conductive oxide (TCO) layer at the front contact, which reduces current leakage. Finally, an anti-reflective coating is applied to the surface. However, cost reduction in CIGS production can be achieved by employing alternative manufacturing processes, such as roll-to-roll processing (R2R). Similar to the normal procedure, R2R processing involves the deposition of a continuous polymer sublayer, over which the succeeding layers are added to form the CIGS. But in this instance, a continuous, flexible substrate is used. Large-scale CIGS production is made possible by this method, which yields a strip of flexible, series-connected cells that, depending on encapsulation, can be combined into rigid or flexible modules. This approach opens up numerous possibilities for various applications (Wang et al. 2019).

The current efficiency record for CIGS solar cells stands at 23.4%, achieved by Helmholtz-Zentrum Berlin. This value is competitive with the best silicon (Si) cell efficiencies. It is important to remember, too, that because large-scale processing presents some difficulties, research cell efficiencies might need to be more easily translated into industrially attainable efficiencies. However, module efficiencies of 20% have previously been attained. In recent years, the efficiency of CIGS cells has exhibited remarkable progress, and further enhancements are anticipated, with ongoing investigations into various areas, such as post-deposition Alkali treatment.

1.2.3.2 Cadmium Telluride (CdTe)

Cadmium Telluride (CdTe) is a promising material for solar energy conversion due to its direct bandgap nature and energy value of 1.5 eV. The first prototype of a CdTe cell was developed in 1972 by Bonnet and Rabenhorst, which achieved a 6% efficiency. Since then, the efficiency of CdTe cells has significantly increased, with a current research record of 22.1% by First Solar in 2015. However, efficiency improvement has not been consistent and attributed to various fabrication breakthroughs. Since CdTe makes up the majority of thin-film modules and has experienced rapid expansion in worldwide production in recent years, the technology is still appealing. A major advantage of CdTe cells is that cadmium, a vital component of the material, is a by-product of zinc refinement, thereby providing an opportunity to repurpose existing waste into renewable energy production. However, there are serious safety and environmental risks associated with the use of cadmium in manufacturing. Because of its long-lasting toxic effects and proven carcinogenic and mutagenic qualities, cadmium is an exceptionally dangerous heavy metal that can destroy aquatic life. Additionally, although not as poisonous as cadmium, tellurium, another component of CdTe cells, poses a risk to aquatic life and can be lethal if breathed (Fraunhofer ISE 2021).

There are two primary configurations for CdTe cells: substrate and superstrate. The first layer functions as the front contact when the layers are placed in the superstrate configuration, usually over a transparent substrate like soda-lime glass. In contrast, the substrate configuration has an inverted layer sequence, allowing for using opaque polymers or metal foil as the base substrate. This characteristic makes it a promising candidate for flexible CdTe cell development, as finding suitable conductive transparent materials for encapsulation is challenging. However, the superstrate configuration currently boasts the highest efficiencies. One of the primary limitations of CdTe cells is the open-circuit voltage, which reduces the efficiency of the cell. The current has significantly decreased despite some attempts to improve voltage through phosphorous doping, resulting in a plateau of total efficiency (Fig. 1.4). Still, if the present loss can be adequately offset, this strategy might be a viable way to increase efficiency down the road. There is a lot of space for improvement because the single-crystalline CdTe theoretical efficiency limit is just about 33% (Burst et al. 2016).

CdTe cells also have the potential to be fabricated in ultrathin dimensions without compromising their spectral efficiency, as demonstrated by their absorber thickness being reduced to approximately 1 μm with negligible efficiency losses. Due to their low weight, this feature makes them particularly attractive for flexible applications and building-integrated photovoltaics (BIPV). Transparent encapsulation materials can be used to create clear CdTe photovoltaic panels with a range of transparency,

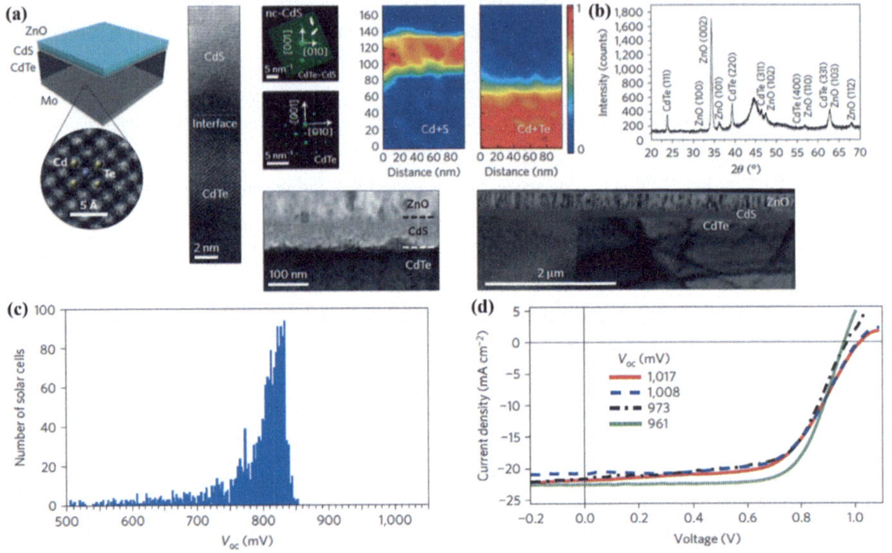

Fig. 1.4 **a** Schematics, HRTEM, TEM, and SEAD patterns of the lattice-mismatched interface of the CdTe solar cell, **b** Grazing Index (@0.25°) XRD patterns of complete device, **c** V_{OC} histogram of over 2400 CdTe solar cells during last decade (NERL), and **d** I–V analysis of CdTe solar cell structured as CdTe/CdS/i-ZnO/ZnO. Reproduced from Ref. Burst et al. (2016), Copyright (2016) with permission of Springer Nature

usually between 10 and 50%. However, it is important to note that increasing transparency inevitably reduces efficiency. Because of their double glass encasement, transparent CdTe panels could be used in place of window panels in buildings to generate power for self-sufficiency and aid in thermal isolation and noise reduction. In addition, transparent CdTe cells may be a promising solution in situations where space is limited, such as on an island (Melo et al. 2020).

1.2.4 Third-Generation Solar Cells (Hybrid Solar Cell)

1.2.4.1 Multi-junction Solar Cells

The inherent characteristics of the semiconductor material employed in PV cells, such as carrier recombination velocity and bandgap energy, impose restrictions on the device's functionality. For instance, silicon, the most widely employed material in photovoltaic cells, has a bandgap energy of approximately 1.11 eV. Therefore, only incident photons with energy equal to or higher than this value will generate charge carriers. However, even in this case, there are still losses due to thermalization and carrier recombination. This limitation becomes evident because only a portion of the solar spectrum has sufficient energy to produce charge carriers in a given semiconductor material. This limit, known as the Shockley-Queisser limit, is estimated to be 29.4% for silicon-based homo-junction solar cells. Various strategies can be pursued to increase efficiency and reduce thermalization losses. One possible strategy is to use a certain material to selectively modify the incident radiation. For example, you might use a prism to match each band with the right substrate. Though it is utilized in some concentrated PV systems, such as photovoltaic mirrors, spectrum-splitting devices, and dichroic mirrors, this technique is not widely used or thought to be viable (Liu et al. 2020b).

A popular method for improving solar cell performance is the multi-junction or tandem construction. This architecture involves stacking multiple semiconductor substrates from top to bottom with decreasing bandgap energies. Photons that enter the cell structure are selectively absorbed by the various layers, with the top layers absorbing higher energy photons and the bottom levels absorbing lower energy photons. This bandgap-selective absorption approach helps to reduce thermalization losses and enhance the efficiency of the cell (Fig. 1.5). By using this architecture, it is possible to surpass the previously established limit known as the Shockley-Queisser limit. The theoretical maximum efficiency value of this architecture is about 85% under concentration for an infinite number of junctions with properly coupled substrates, and about 65% under typical conditions with standard illumination (Geisz et al. 2020).

In the production of multi-junction cells, the lattice constant of each material layer plays a crucial role in addition to the bandgap energy. Dislocations in the structure can result from a stack of materials with varying lattice constants, which can impair the cell's efficiency by causing parasitic losses. Compositionally graded buffers (CGBs)

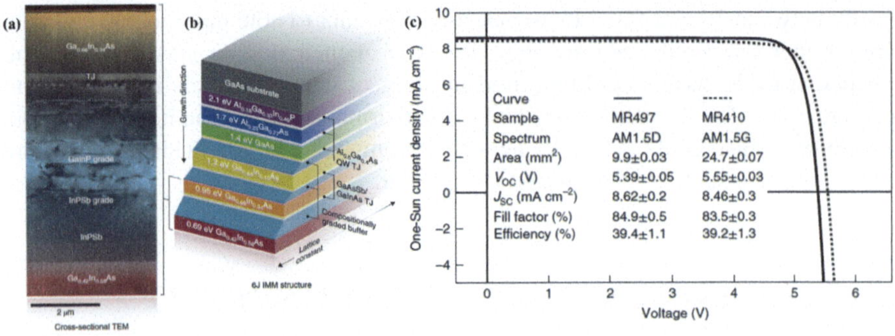

Fig. 1.5 **a, b** Schematic and cross-sectional view of 6-junction inverted metamorphic multi-junction solar cell, and **c** *J*–*V* analysis of 6 J IMM solar cells designed for direct (solid) and global (dashed) spectra at the illumination of 1 Sun. Reproduced from Ref. Geisz et al. (2020), Copyright (2020) with permission of Springer Nature

are inserted between mismatched layers to cushion these consequences and smooth the transition. The buffer layer's chemistry is adjusted along its thickness to more closely align with the lattice constant of the two junctions at its extremities. But there can still be little dislocations. However, the production of lattice-mismatched devices is a more challenging fabrication process and compromises the quality of the material. It is a flexible approach to achieving desired bandgaps and allowing a wider range of materials to be used. Therefore, while choosing materials with matching lattice constants is crucial, introducing lattice-mismatched junctions opens up more possibilities for achieving maximum yield in multi-junction cells.

In the field of semiconductor device engineering, the nomenclature used for multi-junction devices depends on whether the layers have the same lattice constant or not. A structure is considered to be lattice-matched when all layers have the same lattice constant. Conversely, a structure is considered to be lattice mismatched or metamorphic when the layers have different lattice constants. Stacking structures grown monolithically on a substrate with tunnel junctions connecting the layers in sequence make up most multi-junction devices. Depositing the layers with the largest bandgap energies first, then layers with progressively decreasing energy is the favored fabrication process. This mechanism, referred to as "inverted growth," is especially useful in cases where an inverted metamorphic multi-junction device is produced, meaning that the top layers are lattice-matched, but the lower layers are not. When a lattice-matched structure is present, the layers are placed on a narrow, low-bandgap energy substrate that functions as the bottom junction. The device can be completed in fewer steps with this method. But in the event of the inverted arrangement, before removing the substrate on which the structure is formed, a handle needs to be glued to provide structural support. The removed substrate may occasionally be recycled or utilized again, which reduces the additional costs even if it is added.

The choice of fabrication process for multi-junction devices is closely linked to the desired terminal configuration, with two-terminal and four-terminal designs being the main options. The two-terminal configuration is typically achieved through the monolithic growth of all layers in series. In contrast, in order to obtain four exterior terminals, two from each sub-cell the four-terminal layout calls for a mechanically stacked technique. All layers in the two-terminal arrangement are connected in series, which presents technical difficulties because the junction with the lowest value limits the output current. More difficulties arise throughout the fabrication process since every material has unique requirements to preserve its qualities. Careful manipulation is required during the monolithic growth to preserve these properties, making the overall process laborious (Werner et al. 2018).

Conversely, the four-terminal configuration offers several advantages over the two-terminal configuration. The top and bottom sub-cells can be grown separately and operated independently, reducing the cell's sensitivity to spectral variations. This is not possible in the two-terminal configuration. However, the four-terminal configuration still has some drawbacks, including the parasitic absorption losses introduced by transparent bonds and additional transparent contacts. The added freedom provided by the extra contacts also necessitates the use of control electronics, which adds to the overall cost of the device. Overall, the choice of terminal configuration depends on the specific application and performance requirements. Assuming that the optical properties of the sub-cells are optimized, the two-terminal arrangement may still be more efficient even though the four-terminal structure has advantages in terms of spectrum variation and autonomous operation.

1.2.4.2 Organic Solar Cells

One of the most significant challenges of the twenty-first century is the reduction of net carbon emissions toward net zero to slow down the effects of global warming. To achieve this, it will be crucial to implement efficiency measures and transition the electricity, transport, and heating sectors away from fossil fuels toward sustainable technologies as quickly as possible to meet the 1.5 °C target. This transition will require the electrification of significant parts of these sectors, necessitating the large-scale deployment of renewable energy technologies, particularly wind and solar. Crystalline-silicon photovoltaics have experienced rapid growth and currently dominate the solar market, accounting for more than 95% of the sold solar modules. However, several other technologies are under development, with the potential to deliver electricity at even lower cost and accelerate the transition. Adopting these new technologies could significantly reduce carbon emissions, mitigate the effects of climate change, and achieve a sustainable energy future (Philipps et al. 2023).

Organic solar cells (OSCs) are an emerging thin-film photovoltaic technology with the potential to provide a low-cost alternative to traditional solar cells. With advancements in the synthesis of low bandgap organic materials and device processing technology, the PCE of OSCs has surpassed 16% for single junctions and 17% for organic-organic tandem solar cells. Despite these achievements, the commercial use of OSCs could be improved by the poor stability of the devices. Stability is a critical aspect of photovoltaic technology, as it determines the long-term reliability and durability of the devices under various operating conditions. Therefore, developing stable OSCs remains a significant challenge for achieving widespread adoption of this promising technology.

OSCs are a promising renewable energy technology with several advantages, including low cost, lightweight, flexibility, semitransparency, and easy fabrication. Researchers have focused on improving the PCE of OSCs using various strategies, such as designing high-performance donor and acceptor materials, employing ternary active layers, utilizing tandem device structures, optimizing carrier transport layers, and optimizing fabrication processes. As a result, the PCE of OSCs has reached over 16% for single-junction devices and over 17% for ternary and tandem devices. Despite the remarkable progress in PCE, the relatively low stability of OSCs is a significant challenge that hinders their future commercialization. OSCs are prone to degradation caused by various factors, resulting in rapid performance deterioration during operation. Achieving long-term stability in OSCs remains a significant challenge and a critical area of focus for advancing this technology.

OSCs are currently the most promising technology for commercialization. OSCs employ carbon-based semiconductors as the active layer for converting sunlight into electricity. These semiconductors provide a number of benefits, including adjustable electrical characteristics, a large supply of safe raw materials, and the ability to function with low-cost manufacturing techniques like vacuum coating and solution processing. Moreover, OSCs have a low material consumption of approximately 1 g of organic semiconductor per square meter, which, combined with their ability to be processed at low temperatures and on flexible substrates, makes them suitable for large-scale, roll-to-roll production of lightweight devices with a wide range of applications (Abbel et al. 2018). The potential for OSCs to become the cheapest source of electricity is a compelling prospect. The main operational distinction between silicon solar cells and OSCs is that light absorption in organic semiconductor thin films generates bound excitons with limited diffusion length rather than the efficient generation of free charge carriers as in silicon solar cells. This difference has caused OSCs to lag behind silicon solar cells in commercialization. However, in 1986, Tang proposed a solution by achieving efficient exciton separation at the planar interface between two different organic semiconductors—an electron donor and an electron acceptor—resulting in a type II heterojunction (Tang et al. 1986). In 1991, Hiramoto et al. introduced the critical concept of co-evaporating donor and acceptor molecules. This led to a bulk heterojunction (BHJ) with a distributed donor–acceptor interface, overcoming the limited exciton diffusion length. BHJs are now central to all efficient OSCs, regardless of whether they are produced through vacuum or solution-based processes (Hiramoto et al. 1991). A typical OSC stack structure and a monolithic series connection of sub-cells into a module are shown in Fig. 1.6 (Riede et al. 2021).

The continuous advancement of research and development in organic semiconductors tailored for OSC, processing techniques, and stack design has resulted in the development of materials with enhanced absorption and donor–acceptor energy offsets. These advancements have also led to the optimization of BHJ microstructure and stack design, resulting in significant progress in PCEs of around 18% in

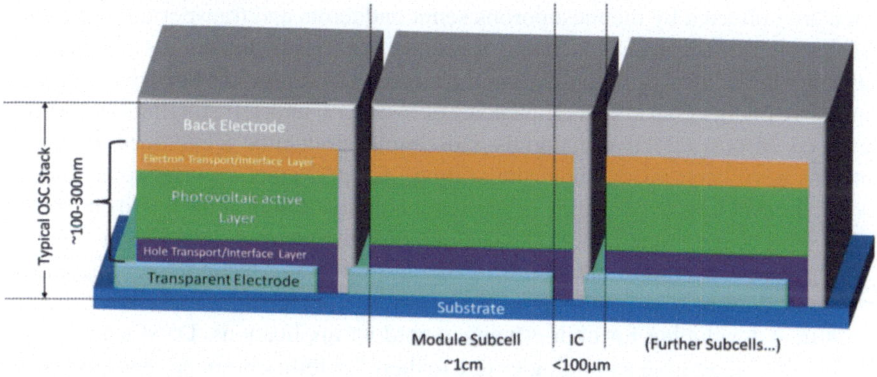

Fig. 1.6 Schematic stack structures of an Organic Solar Cell (OSC) are depicted on the left, while the common monolithic serial connection for an OSC module is shown, with typical dimensions of the module sub-cells and the serial interconnect (IC) indicated. Reproduced from Ref. Riede et al. (2021), Copyright (2021) with permission of WILEY-VCH Verlag GmbH & Co. KGaA, Weinheim

single-junction solar cells. Moreover, PCEs exceeding 20% are becoming increasingly attainable, and module efficiencies are catching up to these values. While PCEs are an essential factor in OSC device and module performance, they alone are insufficient for OSCs to contribute meaningfully to mitigating climate change at scale. Achieving scalability to terawatts of installed capacity and sufficient lifetime at a competitive cost are also necessary. In the following sections, we outline some of the primary research challenges, discuss OSC markets, and provide an outlook on the transformative potential of OSCs concerning cost and carbon emissions (Liu et al. 2020a).

OSCs offer significant potential for innovative photovoltaic applications where their advantageous properties, including low weight, flexibility, and transparency, can be effectively utilized. In the long run, OSCs have the potential to achieve efficiencies and cost competitiveness that would make them suitable for deployment in the mainstream PV power market.

1.2.4.3 Dye-Sensitized Solar Cells

In 1991, Michael Grätzel and his research group developed a new type of solar cell called the dye-sensitized solar cell (DSSC) or Grätzel cell, which offers a promising alternative to traditional inorganic p–n junction solar cells (O'Regan et al. 1991). The DSSC technology combines molecular systems and nanoparticles to create a device that mimics photosynthesis and aims to convert sunlight into a renewable, reliable, and cost-effective energy source. While the idea of injecting dyes into a single-crystal semiconductor was first demonstrated by Gerischer in 1966, it was Grätzel's introduction of a mesoporous semiconductor layer that revolutionized the DSSC technology (Gerischer et al. 1990). The dyes in DSSCs absorb light and generate free electrons, which are collected by the mesoporous semiconductors and transported to the electrode as electrical current. A charge transport material facilitates the regeneration of the dye molecules, thus completing the circuit. DSSC devices have demonstrated impressive energy conversion efficiencies of over 13% under 1-Sun illumination and are based on cost-effective materials and simple production techniques. However, concerns have been raised about the sealing of liquid junction solar cells, and the development of improved sealing strategies or the substitution of the liquid electrolyte with a solid charge transfer material is crucial for their commercialization.

Despite concerns surrounding third-generation solar cell technologies, there must be a clear dominant technology for mass production. Thus, DSSCs are expected to continue developing for high-efficiency outdoor applications. DSSCs are unique among third-generation technologies due to their flexibility in shape, size, and color, making them suitable for various deployment scenarios. They remain a competitive alternative due to their simple preparation methods, lack of toxic materials, and design flexibility. DSSCs can be applied in various environments, from transparent smart windows to consumer electronics and indoor applications, which can power the next digital revolution of widely distributed sensors forming the Internet of Things (IoT).

The irradiance of sunlight is defined as the amount of energy of a certain light wavelength that falls on a unit area per unit of time, measured in J s^{-1} m^{-2} nm^{-1} (W m^{-2} nm^{-1}) and shown in Fig. 1.7a. The overall irradiance can be calculated by integrating the spectral irradiance over all wavelengths in W m^{-2}. DSSCs exhibit excellent performance under sunlight but perform even better when illuminated by artificial light sources that emit spectra similar to visible sunlight. However, indoor lighting conditions differ significantly from outdoor solar irradiance, as the intensity of typical indoor lighting is much lower, and the spectra between different light sources vary widely. Although DSSCs demonstrate commendable performance under sunlight, their efficiency is further enhanced when exposed to artificial light sources (Fig. 1.7b) (Nick84 Wikimedia Commons 2023; Michaels et al. 2021). This is because the absorption profiles of dyes typically used in DSSCs are primarily limited to the visible portion of the solar spectrum. Consequently, when illuminated by artificial light sources with emission spectra resembling the visible range of sunlight, DSCs exhibit even higher-performance capabilities. Illuminance, measured in lx (lumen per unit area, lm m^{-2}), cannot be directly converted to irradiance, as it describes light intensity in terms of human eye perception rather than energy. The spectral properties of the light source must be taken into account to determine irradiance from illuminance. A mathematical equation (Eq. 1.1) can be used to determine illuminance from irradiance once the spectral properties of the light source have been determined.

$$\mathrm{IL}[1x] = \frac{683.002}{A}\left[1\,\mathrm{m}\,\mathrm{W}^{-1}\mathrm{m}^{-2}\right]\int I(\lambda)\cdot E(\lambda)\left[\mathrm{Js}^{-1}\mathrm{nm}^{-1}\right]\cdot \bar{y}(\lambda)\cdot \mathrm{d}\lambda \tag{1.1}$$

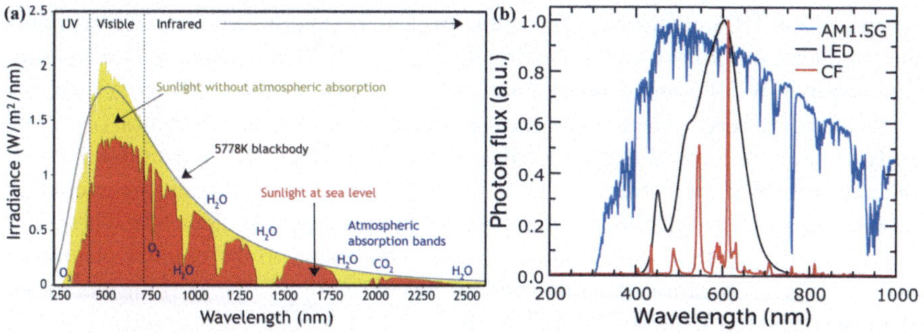

Fig. 1.7 **a** Whole solar irradiation spectrum corresponds to UV-, Visible-, and IR-regions, Reproduced from Ref. Nick84 Wikimedia Commons (2023), copyright (2023) with permission of Wikimedia Commons. **b** Normalized solar spectra under AM1.5G, LED, and white fluorescent standards, Reproduced from Ref. Michaels et al. (2021), copyright (2021) with permission of the Royal Society of Chemistry

The illuminance (IL) and irradiance ($I \cdot E$) can be related by the photopic luminosity function of the human eye, which has a peak response at a wavelength of approximately 555 nm. The irradiance can be obtained by multiplying the light intensity (I) and photon energy (E) over the area (A) of interest. The photopic luminosity function ($\overline{y}$) is dimensionless and integrates the irradiance over all wavelengths, thereby measuring the total light intensity perceived by the human eye.

The scientific community has established several guidelines, exemplified by ASTM standard E948, that prescribe the experimental conditions and testing methodologies to assess the performance of solar cells when subjected to solar irradiation. Nevertheless, no universally recognized protocols have been established to govern the process of indoor measurements.

The DSSC is composed of three fundamental components: the dye-sensitized semiconductor electrode, which is referred to as the working electrode or photoanode, the redox electrolyte, and the counter electrode. A monolayer of dye molecules is adsorbed on the surface of the semiconductor electrode and is responsible for the light absorption mechanism within the device. In the conventional DSSC architecture, the semiconductor electrode displays n-type characteristics, whereby the material's conductivity is attributed to electrons within the conduction band. Furthermore, the semiconductor possesses a wide bandgap and exhibits a negligible contribution to the absorption of solar radiation. Of the numerous semiconductors available for DSSC applications, titanium dioxide (TiO_2) with the anatase crystal structure is the most commonly utilized material. TiO_2 has a bandgap of approximately 3.2 eV and predominantly absorbs ultraviolet light. It should be noted that various alternative semiconductors are viable for use in DSSCs. The absorption of solar radiation by a flat and dense TiO_2 electrode, upon which a monolayer of dye molecules are adsorbed, needs to be improved to achieve significant solar-to-electric conversion efficiencies. To overcome this limitation, high-surface area TiO_2 electrodes are employed, such as mesoporous TiO_2 electrodes. These electrodes are comprised of numerous interconnected nanoparticles with typical dimensions of 20–30 nm, resulting in a porosity of approximately 50% and a surface area that can be several hundred times greater than the projected area. The mesoporous structure facilitates the adsorption of significantly greater amounts of dye molecules than a flat surface. The most effective dye molecules for DSSC performance are chemically bound to the TiO_2 and are in contact with the redox electrolyte that fills the pores of the mesoporous electrode. The redox mediator facilitates the transportation of positive charges to the counter electrode, which is generally positioned parallel and in close proximity to the working electrode.

The primary step in the operational mechanism of a DSSC, as illustrated in Fig. 1.8, is the photoinduced electron transfer from the dye molecule to the conduction band of the TiO_2 semiconductor. Upon the absorption of light by the dye molecule (S), an electron is excited to a higher energy level, generating an excited state dye molecule (S^*). This excited dye molecule can subsequently transfer an electron into the conduction band of the TiO_2 semiconductor, which comprises a range of acceptor levels. The timescale of this electron transfer process is on the order of femto- to pico-seconds. In a mesoporous semiconductor, electrons are balanced by ions from the electrolyte surrounding it, and their movement is primarily driven by electronic drift–diffusion. When subjected to full sunlight, electrons are collected at the electrode contact within a millisecond time scale. However, the slow and light-dependent electron transport is typically explained using a multiple-trapping model that posits an exponential distribution of traps beneath the conduction band. Despite this model being widely accepted, the precise nature of the traps remains a topic of ongoing debate.

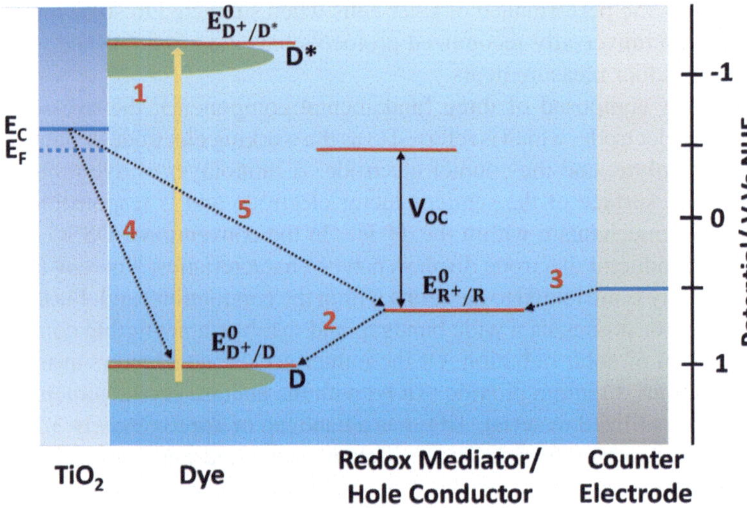

Fig. 1.8 Schematic display of a DSSC device and its tentative working mechanism

Recent research suggests that cations adsorb onto the surface of mesoporous TiO_2 upon electron accumulation. This accumulation, in turn, could result in electrostatic traps for the electrons within the semiconductor. This new finding may explain why similar trap distributions are observed across various metal oxides.

In a DSSC, the sensitized TiO_2 is in contact with an electrolyte that contains a redox mediator (R^+/R or I_3^-/I^-) that facilitates the regeneration of the dye and also enables the transfer of positive charges from the working to the counter electrode through diffusion of R^+. At the counter electrode, R^+ is reduced to R. The dye regeneration process occurs on a microsecond time scale and must be sufficiently fast to prevent the recombination of electrons from the semiconductor to the oxidized dye. Electrons can also recombine with the oxidized form of the redox mediator. The energy level diagram of the DSSC is illustrated in Fig. 1.8 (Srivastava et al. 2022). The ground state energy level of the dye is located just below $E^0(S^+/S)$, the standard reduction potential of the dye, and is referred to as the highest occupied molecular orbital (HOMO) level. The energy level of the excited dye S^* is obtained by adding the absorbed photon energy. The lowest-lying excited state level is obtained by adding $E_{0\text{–}0}$, typically obtained experimentally from the intercept of normalized absorption and fluorescence spectra, and is known as the lowest unoccupied molecular orbital (LUMO) level (Srivastava et al. 2022).

In a DSSC, the energy level alignment between the dye and the semiconductor is critical for efficient photoinduced electron injection, as S^* levels need to be higher than the conduction band edge EC of the semiconductor. However, non-radiative decay and fluorescence of the dye can compete with the injection reaction. To optimize DSSC performance, a good match between the energy levels of the dye and redox mediator is required to ensure sufficient driving force for the reduction of the oxidized dye and prevent losses through electron/dye recombination. The voltage output of the DSSC is determined by the potential difference between the working and counter electrodes, where the potential of the counter electrode is close to that of the redox potential of the electrolyte, and the potential of the working electrode is equal to the Fermi level (E_F) of the semiconductor and given by Eq. 1.2:

$$E_F = E_C - \frac{k_B T}{e} \ln \frac{n_c}{N_c}. \tag{1.2}$$

The relationship between the density of conduction band electrons (n_c), the effective density of electronic states at the bottom of the conduction band (N_c), and temperature (T) is described by the Boltzmann constant (k_B) and elementary charge (e). At room temperature, the ratio of k_B*T/e is 0.0257 V. For the TiO_2 anatase phase, N_c is approximately 10^{20} cm^{-3}. To achieve a high output voltage in solar cell operation, it is necessary to maintain a high density of conduction band electrons, which can be achieved by minimizing the rate of electron recombination. This results in a Fermi level closer to the conduction band, which leads to higher voltage output.

Despite significant progress, there are still knowledge gaps in understanding the DSSC. Charge recombination is the primary cause of efficiency loss in DSSCs and other solar cells. Altering any of the DSSC components, including the dye, redox shuttle, or semiconductor, can significantly impact multiple processes, either positively or negatively. Consequently, any modifications to DSSC materials must consider the impact on the overall system performance. DSSCs are intricate devices, and enhancing the efficiency or stability of a single component does not guarantee that the desired targets will be achieved for the device as a whole.

1.2.4.4 Quantum Dot Solar Cells

Quantum dots (QDs), once a specialized field, have become a viable material for various commercial optoelectronic devices. This article focuses on advancements in lead-based QD solar cells (QDSCs) from the perspective of pathways for the relaxation of excited states to the ground state. A systematic understanding of the fundamental processes in QDs has significantly improved solar cell efficiency from approximately 3% to over 13% within eight years. We have compiled data from about 200 articles that report functional QDSCs to provide an overview of the current limitations of the technology. Our findings indicate that the open-circuit voltage significantly limits the device's efficiency, and we propose various strategies to overcome this limitation.

Colloidal semiconductor quantum dots (QDs) are nanocrystals suspended in a solution. Due to their solution processability, adjustable properties, and unique photo physics, they have gained significant interest in optoelectronic devices, imaging, and fundamental research. The size and size distribution of the QD ensemble have a direct impact on the material's energy levels, including the HOMO to LUMO gap or bandgap. This phenomenon can be explained from two perspectives. In the bottom-up viewpoint, the molecular orbitals combine with each additional atom bonded to the QD, gradually narrowing the HOMO–LUMO gap until the material reaches the "bulk" phase. The second top-down perspective involves spatially confining the exciton's wave function (an electron–hole pair) by reducing the crystal's size, which increases the energy needed to excite the QD, similar to the particle-in-a-box concept (Efros et al. 1998).

Various methods exist to synthesize QDs, as outlined in various literature. The most common approach, known as hot injection, follows a nucleation and growth reaction pathway similar to La Mer's model. Following injection, the precursor concentration in solution surpasses the nucleation threshold, resulting in a burst of nucleation and subsequent QD growth. This process continues until the precursor concentration falls below the nucleation threshold, and the growth regime begins. During growth, size focusing may occur if the QD growth rate is size-dependent, leading to a narrower size distribution. However, the high-surface energy of small QDs can drive their dissolution and precipitation onto larger QDs, resulting in Ostwald's ripening and an increase in size distribution. By modifying reaction temperature, precursor reactivity, and/or reaction kinetics, these processes can be manipulated to control QD size and size distribution. The growth regime ends when all reactants are consumed and the concentration approaches zero. Surface chemistry and environment significantly affect many QD properties due to the high proportion of atoms exposed on the surface compared to those in the core, not just the size of the QDs (Ganesan et al. 2018).

Herein, the photoexcitation processes in colloidal semiconductor QD films and devices were discussed, considering both the size and surface effects. We provide a comprehensive overview of the relevant processes occurring within different time scales, ranging from a few femtoseconds to microseconds. We subsequently provide a detailed analysis of the advancements made in Pb-based QDSCs. Readers interested in further details on specific subtopics should consult more specialized reviews. The following sections aim to explore the various processes that occur when a QD absorbs a photon. By delving into the details of these processes and their corresponding time scales, as depicted in Fig. 1.9, it becomes possible to design optoelectronic devices more strategically (Ganesan et al. 2018). Furthermore, we will showcase device studies that apply the knowledge gained from each process, providing an overview of the history and current status of Pb-based QDSCs. The initial and crucial step toward using QD materials in optoelectronics is the propagation and absorption of light. As previously discussed in works by Leatherdale et al. and Ricard, the absorption cross section can be expressed mathematically as a function of several parameters, including the angular frequency of light (ω), the speed of light (c), the complex refractive index ($n'_s = n_s - ik_s$) of the solvent or surrounding medium (designated by the subscript s), the absorption coefficient of the bulk material ($2n_1k_1 = \alpha$), the radius of the particles (a), the extinction coefficient (k), and the local field factor $f(\omega)$ (Eqs. 1.3 and 1.4).

$$\sigma = \frac{\omega}{n_s'^2 c}|f(\omega)|^2 2n_1k_1\left(\frac{4}{3}\pi a^3\right) \tag{1.3}$$

$$f(\omega) = \frac{3n_s'^2}{n_1'^2 + 2n_s'^2} \tag{1.4}$$

Light absorption in a film of QDs can be increased by manipulating specific properties, enhancing a solar cell's overall electrical current and power output. However, for optimal efficiency, the thickness of the film must be carefully controlled, as increasing thickness may not necessarily lead to increased light absorption. Researchers have explored incorporating plasmonic materials with localized surface plasmon resonance (LSPR) into QD solar cells to address this (Luther et al. 2011). This strategy has improved light absorption in the near-infrared (IR) region and other parts of the solar spectrum, depending on the type and position of the plasmonic nanocrystals used. For example, incorporating plasmonic gold nanorods with HgTe QDs or gold nanodisk arrays with HgSe and HgTe QDs has enhanced photocurrent extraction and improved the optoelectronic properties of the QDSCs (Chen et al. 2014).

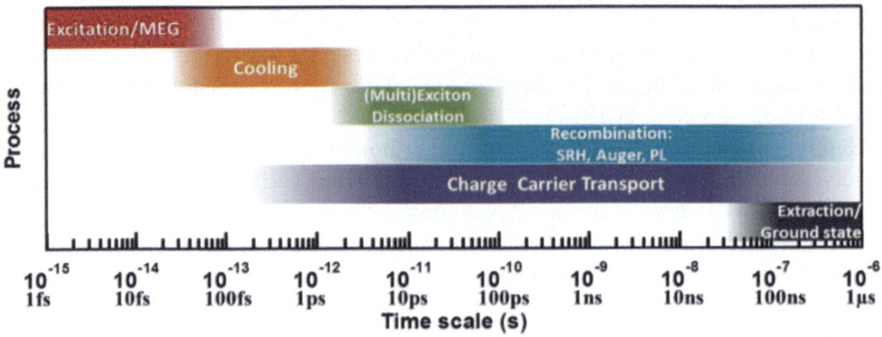

Fig. 1.9 Details of different processes in the lifetime of photoexcitation and their corresponding time scales. Reproduced from Ref. Ganesan et al. (2018), Copyright (2018) with permission of MDPI

In solar cells with thin absorber layers due to limited minority carrier diffusion lengths, improving the path length of incident light waves through the absorber material by enabling multiple passes can increase the probability of light absorption, thus improving device performance. However, this method is challenging to apply in QDSCs due to the non-uniform accumulation of QD solution on textured metal oxide electrodes during spin-coating or dip-coating processes. Nevertheless, Sargent's research group has developed optimized texturing techniques to overcome this challenge and achieve high-performance QDSCs. Furthermore, Hens et al. (2012) have proposed new equations to express the absorption coefficient μ for a QD dispersion with a volume fraction of f, given as:

$$\mu = \frac{n_1}{n_s} |f(\omega)|^2 f \alpha. \tag{1.5}$$

The absorption coefficient of QDs is significantly influenced by their environment, including their surface ligation of the QDs. This is evident from the observation that adding p-methylbenzenethiolate to a solution of PbS QDs results in a broad increase in absorption and a shift of the first exciton transition peak toward longer wavelengths. This is surprising since p-methylbenzenethiolate has negligible absorption in the relevant spectral region. Furthermore, the increase in the molar absorption coefficient at 400 nm, which is approximately 340%, is not fully explained by the change in $f(\omega)$ predicted by the Maxwell-Garnet practical medium theory. This suggests that the interaction between the ligand and QD system is more complex and significant than the simple sum of their individual contributions (Giansante et al. 2015).

The size of a QD determines the energy of a photon that can be absorbed to create an exciton, and if the absorbed photon's energy exceeds twice the bandgap, multiple decay paths to the ground-state energy levels of the excitons become available, each with its characteristic timescale. In QD semiconductors, an Auger pathway allows for the possibility of an inverse Auger pathway, as depicted in pathway **1(a)** of Fig. 1.10a, in addition to the traditional Auger pathway represented by pathway

2(a). Auger processes are three-body events where one carrier transfers energy to a third carrier by interacting with another carrier of the opposite charge. The photoexcitation of a high-energy photon ($h\nu > 2E_g$) can split a high-energy electron–hole pair into two or more electron–hole pairs of lower energy (Gao et al. 2015). We will discuss the phenomenon of multiple exciton generation (MEG), also known as carrier multiplication (CM), in cases where the excitations yield free charges. This effect was observed in QDs at much lower bandgap multiples than in bulk semiconductors. Meanwhile, CM in bulk semiconductors, known as impact ionization (II), is significant only at photon energies far into the UV range (> 3 eV), which contributes to less than 1% of the solar spectrum.

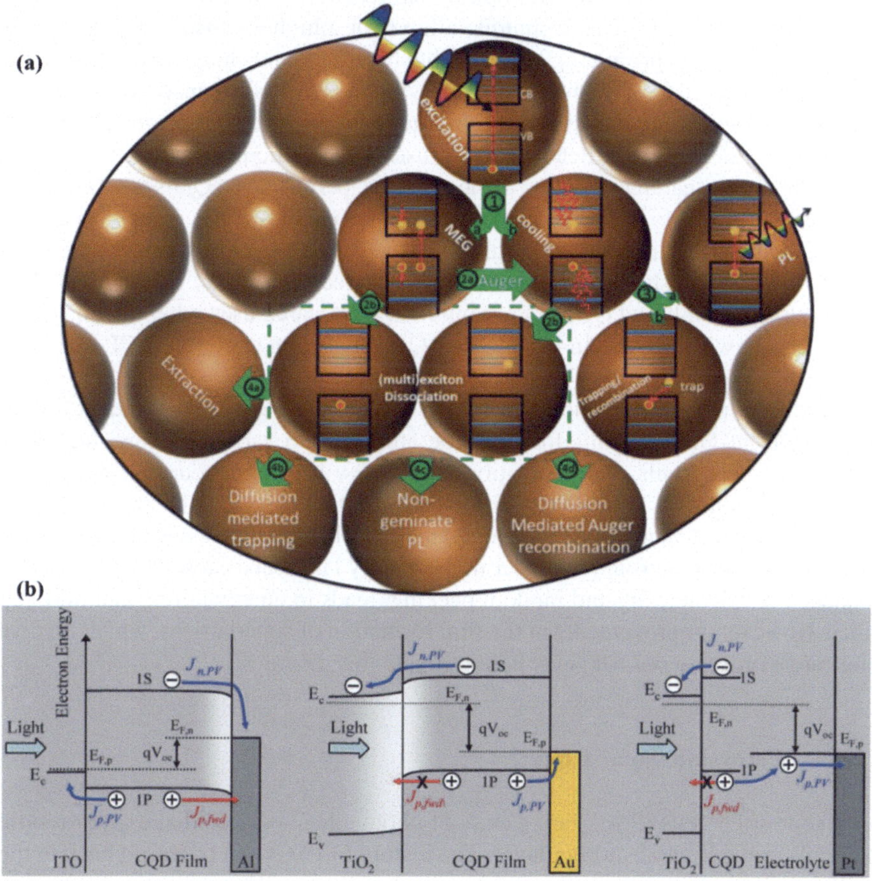

Fig. 1.10 **a** Step-wise schematic depiction of free charge creation in QD films, Reproduced from Ref. Ganesan et al. (2018), Copyright (2018) with permission of MDPI. **b** Schematic band diagram of three most popular PbS QDSCs, Reproduced from Ref. Pattantyus-Abraham et al. (2010), Copyright (2010) with permission of the American Chemical Society

The performance of a QDSC is significantly affected by its design and structure. Figure 1.10b shows three different QDSC structures, the rightmost being QD-sensitized solar cells that were initially fabricated using n-type mesoporous TiO_2 (m-TiO_2) sensitized with PbS QDs (Pattantyus-Abraham et al. 2010). Early research by Hoyer et al. showed photoconduction in a device with a bandgap of $\approx$1.85 eV where the QD energy bands were favorably aligned with the cathode, allowing for efficient carrier extraction from the QDs (Hoyer and Könenkamp 1995). Subsequent research by Ju et al. sandwiched the above setup between tin-doped indium oxide (ITO) and gold, achieving a PCE of 3% (Ju et al. 2010). However, the drawbacks of the mesoporous structure pushed QDSCs toward planar solid-state configurations. Nevertheless, research on sensitized solar cells continued developing a technique known as successive ion layer adsorption and reaction (SILAR) to synthesize PbS QDs in situ on m-TiO_2. This technique allows for a higher density of PbS QDs in the pores of m-TiO_2, promoting higher light absorption and charge extraction at the interface between TiO_2 and PbS QDs, improving device performance. All solid-state sensitized cells without polysulfide electrolytes have also been fabricated using TiO_2 nanorods similar to a bulk heterojunction used by organic photovoltaics. This has opened up new research avenues to achieve higher-performance QD-sensitized cells. Notable achievements include devices with Hg^{2+} doped PbS QDs made through the SILAR method with exceptionally high photocurrents of 30 mA/cm^2 and devices using PbS/CdS core/shell QDs that exhibit PCEs greater than 7%. While this discussion focuses on Pb-based QDs, a growing field of research uses other QD materials as sensitizers, with Zn–Cu–In–Se (ZCISe) QDs achieving over 12% PCE (Yu et al. 2017).

The difference between the bandgap energy and the V_{OC} is a significant limitation of current QDSCs. This loss may be attributed to various factors, including Urbach or energy tails, recombination/trap states, large dark current, and insufficient heterojunction characteristics. Except for two devices, the practical limit for V_{OC} is approximately half the bandgap, regardless of the device configuration, bandgap, or research group. Disorder in the energy landscape caused by film formation, such as QD necking and random packing, leads to energy tailing and mid-gap states. However, improvements in the film formation of superlattices, which involve long-range ordering of QDs, may help mitigate this disorder.

1.2.4.5 Perovskite Solar Cells

Perovskite solar cells (PSCs) are a class of photovoltaic devices that use perovskite materials as the light-absorbing layer. The history of PSCs can be traced back to the 1990s when Tsutomu Miyasaka and his team discovered that perovskite materials could be used as a light-absorbing layer in DSSCs. However, it was not until 2009 that Henry Snaith and his team reported that methylammonium lead iodide ($MAPbI_3$), a perovskite material, could make efficient solid-state solar cells. Since then, PSCs have attracted a lot of attention from researchers worldwide due to their high PCE, low cost, and simple fabrication process. The PCE of PSCs has been increasing rapidly in the last decade, with the record PCE exceeding 25% in 2022, comparable to the best-performing silicon solar cells (Miyasaka et al. 2022).

The rapid progress in PSCs is mainly attributed to discovering new perovskite materials, developing new device architectures, and optimizing fabrication processes. In addition to the traditional $MAPbI_3$ perovskite, other materials such as formamidinium lead trihalide ($FAPbI_3$) and cesium lead trihalide ($CsPbI_3$) have also been explored, which have shown to have better stability and higher PCEs. Researchers have also developed new device architectures, such as tandem and inverted structures, to improve the PCE and stability of PSCs. Tandem structures can combine different perovskite materials with complementary absorption spectra to increase the light-harvesting efficiency. In contrast, inverted structures can reduce the degradation of perovskite materials by preventing the diffusion of moisture and oxygen. The fabrication process of PSCs has also been optimized by introducing new methods, such as anti-solvent dripping, blade coating, and spray coating, which can produce high-quality perovskite films with good coverage and uniformity.

Despite the remarkable progress, PSCs still face some challenges, such as their stability under ambient conditions and lead toxicity in the perovskite materials. Nevertheless, the rapid advancement in PSCs has opened up new possibilities for low-cost, high-performance, and scalable solar cells, making them one of the most promising candidates for future energy generation. One of the major challenges is stability, as PSCs are highly sensitive to moisture, heat, and light. These factors can cause degradation of the perovskite material, leading to decreased efficiency and, eventually, device failure. To address this challenge, researchers are exploring various strategies such as encapsulation, using additives, and developing more stable perovskite materials. Another challenge is the toxicity of lead-based perovskites, which poses potential environmental and health risks. To overcome this issue, researchers are exploring the use of lead-free perovskites, such as tin-based perovskites, which have shown promising results in terms of efficiency and stability (Vidyasagar et al. 2018).

Engineering challenges include the development of efficient and cost-effective large-scale manufacturing processes for PSCs, as well as the integration of PSCs with other components such as batteries and electrical grids. Additionally, the use of transparent conductive electrodes and the optimization of charge transport in the device architecture are areas of ongoing research. Despite these challenges, the rapid progress of PSCs over the past decade has shown their potential to become a key player in the renewable energy landscape. With continued research and development, PSCs have the potential to provide a low-cost and highly efficient source of renewable energy for a range of applications.

PSCs have shown remarkable progress in efficiency and stability over the past few years, making them a promising candidate for next-generation photovoltaics. Some of the potential future applications of perovskite solar cells include:

1. Building-integrated photovoltaics (BIPV): Perovskite solar cells can be integrated into building materials such as windows, roofs, and facades, enabling them to generate electricity while also serving as a part of the building's design.
2. Portable and wearable electronics: Perovskite solar cells' flexibility and lightweight nature make them suitable for use in portable and wearable electronics such as smartphones, smartwatches, and fitness trackers.
3. Indoor photovoltaics: Perovskite solar cells have shown the potential to generate electricity under low light conditions, making them suitable for use in indoor environments such as offices, homes, and shopping malls.
4. Large-scale solar farms: The low cost and high efficiency of perovskite solar cells make them a potential candidate for large-scale solar farms, which can provide renewable energy to power cities and industries.
5. Tandem solar cells: Perovskite solar cells can be combined with other types of solar cells, such as silicon, to create tandem solar cells, which have the potential to achieve even higher efficiencies than standalone perovskite solar cells.

Overall, perovskite solar cells hold great promise for a wide range of applications and have the potential to transform the renewable energy landscape. However, some technical and commercial challenges still need to be addressed before becoming mainstream technology, such as improving their long-term stability, reducing production costs, and scaling up production to meet demand.

1.3 Perovskite Solar Cells: A Miracle

PSCs are the main focus of this work because they can achieve extremely high efficiency at a lower cost than other conventional cells. PSCs appear to be a strong contender for achieving high efficiency at minimal resource and manufacturing expenditures. The capacity of perovskite material to interact with various light spectra to maximize the amount of incident solar radiation converted to energy is its greatest advantage over traditional PV. Its ability to be manufactured on a flexible substrate allows for various applications (Chand Yadav et al. 2022). The PSC's advantages include being lightweight, having a customized form factor, being simple to produce, being scalable, and many more. Figure 1.11 depicts the yearly progress made in the performance enhancement of PSCs.

Although becoming a viable candidate, it is still in the early stages of commercialization relative to other PV technologies. The pricing of the device is increased by the fact that the effective PSCs still use Au as an electrode and some other very costly materials. However, investigating the inexpensive electrode can fix this flaw. Since the majority of the most popular PSCs contain Pb, they are poisonous by nature. Even though research is being done to replace the hazardous Pb, all Pb-free PSCs have yet to match the best performance achieved by Pb-based PSCs. The scientific community is aggressively seeking Pb substitutes to solve the toxicity issue. Sn has performed the best out of all the possible replacements. Due to poor perovskite

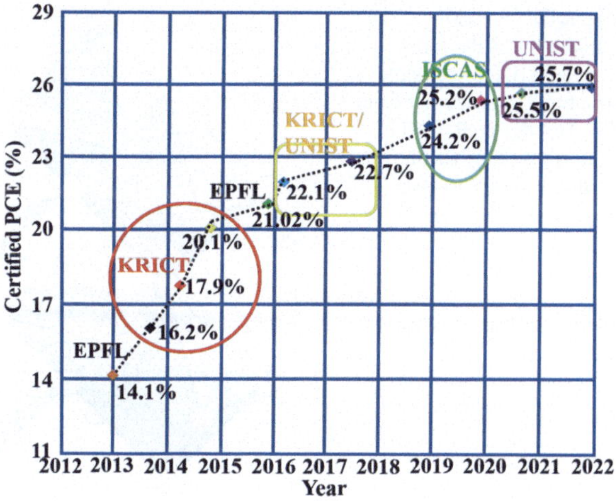

Fig. 1.11 Verified yearly PCEs of the various PSCs obtained by different research groups. All the data has been taken from NREL, 2022

stability and an inadequate energy level discrepancy between the charge transport layer and perovskite, Sn-based PSCs performed less well than Pb-based PSCs (Cao et al. 2021).

The most anticipated new generation of solar cells is perovskite solar cells, which are the topic of this discussion. Investigating perovskite's structure helps pinpoint the essential elements that make it work well in solar cells. The main developments, procedures, problems, and prospects for perovskite solar cell technology are discussed.

1.3.1 Perovskite Structure

Any substance with a similar crystal structure of calcium titanate ($CaTiO_3$) is called "perovskite" (Snaith et al. 2013). Perovskite has a crystalline structure with the formula ABX_3, where A and B are typically monovalent and divalent cations, and X can be oxygen or a halogen anion. Six anions enclose every B-cation to form $[BX_6]^{4-}$ octahedra. The ReO_3 structure is formed by connecting these through corner-sharing (Evans et al. 2020). The A-cation occupies the cuboctahedral position in perovskite structures, as shown in Fig. 1.12 (Hoefler et al. 2017). The capacity to incorporate elements with various valence states into the framework is among the main advantages of the perovskite structure.

For instance, cations A and B would be divalent and trivalent if X is an oxygen anion (O^{2-}). On the other hand, the cations A and B would be monovalent and divalent, respectively, if X were a halogen anion (Cl^-, Br^-, or I^-). This offers a fantastic potential to change the material's properties, like its bandgap, quickly (Park et al. 2018). The ultimate geometry and longevity of perovskites are significantly influenced by these ion charges and sizes utilized in the material. There are three

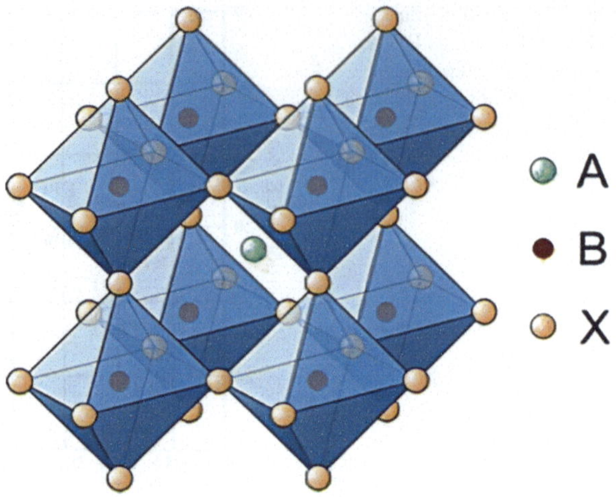

Fig. 1.12 Typical schematics of the most famous ABX_3 perovskite structure. Reproduced from Ref. Hoefler et al. (2017), Copyright (2017) with permission of Springer Nature

rules that are necessarily followed during the metal halide perovskite formation: (i) charge neutrality among A-, B-, and X-sites (Eq. 1.6), (Qian et al. 2016) (ii) octahedral factor (μ) (Eq. 1.7), which can be utilized to predict the stability of BX_6 octahedra, (Travis et al. 2016) and (iii) the most popular and effective ratio is the Goldsmith tolerance factor (t) (Eq. 1.8) (Huang et al. 2016).

$$N(\mathrm{A}) + N(\mathrm{B}) = 3N(\mathrm{X}) \tag{1.6}$$

$$\mu = \frac{r_\mathrm{B}}{r_\mathrm{X}} \tag{1.7}$$

$$t = \frac{r_\mathrm{A} + r_\mathrm{X}}{\sqrt{2}(r_\mathrm{B} + r_\mathrm{X})} \tag{1.8}$$

where N, r_A, r_B, and r_X are defined as the valance and ionic radii of the A-, B-, and X-sites. Ionic size limitations established by the X_6 octahedron constrain the inclusion of the B-site cation. It has been discovered that metal halide perovskites are most stable when μ values are $0.442 \leq \mu \leq 0.895$ (Li et al. 2008). The degree to which the A-cation fits within the perovskite system is effectively measured by the "t". The formation of perovskite structures is only taking place if "t" satisfies the criterion of $0.8 \leq t \leq 1$. In contrast, the value $t = 1$ is defined as the perfect fit for ideal ABX_3 cubic phase perovskite structure formation (Travis et al. 2016). If $t = (0.9–1.0)$, then mostly cubic phase perovskites were formed. However, if $t = (0.8–0.89)$, distorted orthorhombic, tetragonal, or maybe rhombohedral structured perovskites were formed as the ionic radii of A-site continuously decreased. On the other hand, if $t > 1$, i.e., A-site is very large in size, so instead of forming a perovskite structure, it will deform into a hexagonal structure containing stacks of face-sharing octahedra (Uribe et al. 2016; Kieslich et al. 2014).

Table 1.2 Effective ionic radii of various organic, inorganic, metallic cations, and halide anions

Cation A	Effective radii, r_A (pm)	Cation B	Effective radii, r_B (pm)	Anion X	Effective radii, r_X (pm)	References
Ammonium, $[NH_4]^+$	146	Pb^{2+}	119	Iodide, I^-	220	Kieslich et al. (2014) Shannon (1976) Chen et al. (2015) Saparov and Mitzi (2016)
Hydroxylammonium, $[NH_3OH]^+$	216	Sn^{2+}	110	Chloride, Cl^-	181	
Methylammonium, $[CH_3NH_3]^+$	217	Sn^{4+}	69	Fluoride, F^-	129	
Formamidinium, $[CH(NH_2)_2]^+$	253	Ge^{2+}	73	Bromide, Br^-	196	
Hydrazinium, $[NH_3NH_2]^+$	217	Cu^{2+}	73	Formate, $HCOO^-$	136	
Dimethylammonium, $[(CH_3)_2NH_2]^+$	272	Fe^{2+}	78			
Ethylammonium, $[(CH_3CH_2)NH_3]^+$	274	Pd^{2+}	86			
K^+	164	Sb^{3+}	76			
Rb^+	172	Bi^{3+}	103			
Cs^+	188	La^{3+}	103			
		Te^{4+}	97			
		Sm^{3+}				

The ideas of isovalent and aliovalent substitution in metal halide perovskites can also be explained using these replacement principles. To forecast new lead-free perovskite complexes for solar applications based on the ionic radii of the participating ions (Table 1.2), the Goldschmidt substitution rules have received a lot of attention nowadays. Figure 1.13a shows the estimated t and μ values for different perovskite materials. In contrast, Fig. 1.13b depicts the investigated (t, μ) values for relative stability between any two perovskites according to the new factor $(\mu + t)^{\eta}$ (η: atomic packing fraction) demonstrated by Chen et al. (2015) and Sun et al. (2017).

1.3.2 Device Architecture

Generally, PSCs consist of five layers, which are (i) a transparent conducting electrode, an electron transfer layer (ETL), an active perovskite absorber, a hole transfer layer (HTL), and metallic contacts. The PSCs device architecture can be typically categorized into six types: (i) the mesoscopic *n–i–p*, (ii) mesoscopic *p–i–n*, (iii) planar *n–i–p*, (iv) planar *p–i–n*, (v) ETL-free, and (vi) HTL-free configurations. Figure 1.14a–c depicts three widespread device configurations (Srivastava et al.

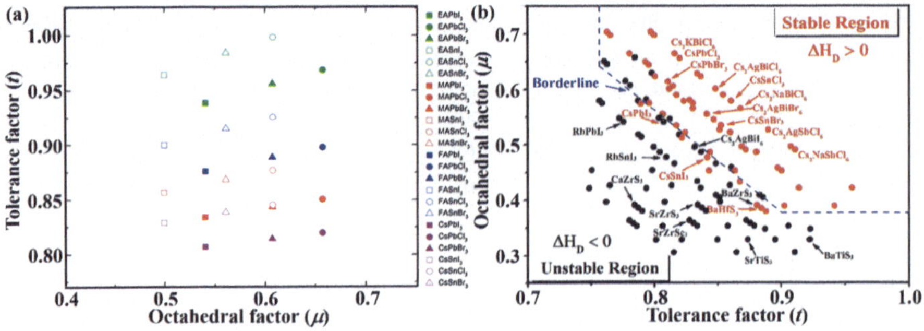

Fig. 1.13 **a** Graphical representation of the estimated octahedral factor (μ) for various PSCs utilizing different organic/inorganic cations, Reproduced from Ref. Chen et al. (2015), Copyright (2015) with permission of Elsevier. **b** Map of (t, μ) for estimating the relative stability of the different PSCs, Reproduced from Ref. Sun et al. (2017), Copyright (2017) with permission of the American Chemical Society

2023b). The transparent conductive electrodes (TCE: FTO or ITO) allow the incident sunlight to pass through and enter the active perovskite materials. After the received radiation is absorbed by the active perovskite absorber, electron, and hole pairs are produced. With the aid of exciton contacts, these produced electrons and holes are carried toward the outer circuit via ETL and HTL, respectively. To guarantee effective exciton collection, the material employed in ETL and HTL must possess the necessary band alignment about the active absorber materials. All three layers (ETLs, HTLs, and active perovskite absorber) must be carefully chosen and assembled to obtain the most outstanding performance out of the PSCs. In this regard, Roy et al. revealed that the optimum thickness of the active perovskite materials must be chosen so that the diffusion lengths of the generated photo charges match the material's thickness (Roy et al. 2022).

For charge transport layers, either organic or inorganic materials are used. The expense and purity requirements for organic HTLs are very high. Additionally, some research implies that organic HTLs continually take part in deterioration, which lowers PSC longevity. A good choice of HTL is essential for the growth of PSCs in the future. The inorganic HTLs are a possible substitute. The inorganic HTLs are more stable, inexpensive, and simple to synthesize. Prior to choosing any specific material, perovskite must match the necessary energy level. P-type GO, C, CuSCN, $CuZnSnS_2$, $CuInS_2$, CuI, Cu_2O, CuO, NiO, and others can be utilized as inorganic HTLs (Elseman et al. 2019). The PCE and device durability are both heavily influenced by the ETL layer. Due to its higher longevity, TiO_2 is the most commonly utilized ETL. Finding a substitute is compelled because it necessitates a higher annealing temperature and promotes the ion immigration process that destroys PC cells. N-type ZnO, CdS, PCBM, SnO_2, and other materials are examples of ETLs. The most often used ETL material is SnO_2, which has a wider bandgap than TiO_2 (3.6–4.1 eV) and a steeper conduction band (CB). Due to its (i) larger mobility of 240 cm^2/V, (ii) steeper CB that improves charge carrier collection and disposal,

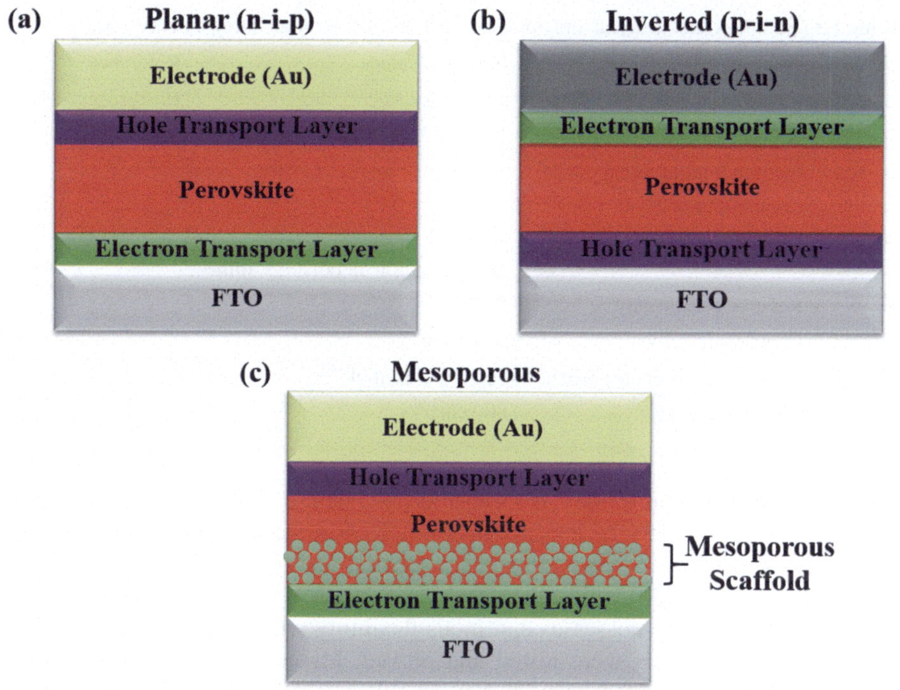

Fig. 1.14 Schematic of three most common PSC device architectures. Reproduced from Ref. Srivastava et al. (2023b), Copyright (2023) with permission of Elsevier

(iii) broader bandgap, (iv) low-temperature manufacturing, and (v) excellent optical transmittance and conductivity, SnO_2 is an attractive choice for ETL (Raj et al. 2021).

A TCE is essential for a solar cell's operation, which directs incident light into the absorber layer and pulls charge carriers produced by photosynthesis for the external circuit. TCEs primary characteristics are appropriate conductivity and transparency, so the balance in both parameters is essential for superior photovoltaics.

1.3.3 Preliminary Features

Understanding the underlying characteristics of the materials is necessary to construct PSCs with the maximum PCE. This comprises the light absorbers' optical and electrical characteristics and how the photoinduced charges are harvested. Perovskite materials have unique qualities that enable them to function as efficient PV devices. The excellent efficiency is attributed to both the long diffusion wavelength (> 1 μm) and the strong electron mobility (800 cm^2/Vs). They exhibit a high absorption coefficient (10^5 cm^{-1}) due to the *s*–*p* antibonding coupling. Small exciton binding energy,

Table 1.3 Summarized electrical and optical properties of commonly known perovskites

Sr. No.	Properties	Expected values	Reference
1	Bandgap	1.5–2.5 eV	Roy et al. (2020)
2	Absorbance coefficient	10^5 cm^{-1}	
3	Exciton binding energy	< 10 MeV	
4	Quantum efficacy	> 70%	
5	Defect density	10^{10} cm^3 in SC 10^{15}–10^{17} cm^2 in polycrystalline	
6	Relative permittivity	3	
7	Charge longevity	> 300 nm	
8	Crystallization energy barrier	56.6–97.3 kJ mol^{-1}	
9	Charge mobility	800 cm^2/Vs	

estimated to be less than 10 meV, is another advantageous feature of perovskite materials (Miyata et al. 2015). As a result, the excited carriers are free to relocate freely. Table 1.3 summarizes these defining characteristics and the values that correspond to them (Roy et al. 2020).

Perovskite structure possesses enormous flexibility with all A-, B-, and X-sites where various materials were tested and proved. Mixed-halide perovskite-based PSCs have emerged as the best solution to enhance longevity and performance, which is remarkably helpful in tandem solar cell fabrication. Various isovalent and aliovalent elements are employed instead of lead (Pb) metal cations to address the toxicity issue. However, the efficiency of lead-free PSCs is still inferior to that of lead-based PSCs. Due to low efficiency, efforts to supplant the Pb cation still need further interest among researchers (Wu et al. 2021). Table 1.4 summarizes a variety of PSC architectures and their reported performances.

1.3.4 Working Principle

The photoelectric effect is the fundamental component that drives a solar cell's operation. When a solar cell is exposed to light, electron–hole pairs are produced. As a result of the local electric field forces at the p–n junction, the holes and electrons proceed in different directions toward the terminals. As a result of this movement, the semiconductor's potential difference rises. Different photovoltaic effects can be categorized according to where the local electric fields originate from.

The p–n junction is the primary photovoltaic effect that is relevant for studying solar cells. It produces an electric field that is only present in a small transition region and is directed from the n to the p side (depicted in gray in Fig. 1.15) with a width of d. A positive potential difference, or voltage, exists between the p and n sides of the junction as a result of the electrons and holes being able to separate due to this electric field. This voltage is observable even in the absence of any electrical

Table 1.4 PCE and other photovoltaic parameters of different device configurations have been reported to date

Device architecture	J_{SC} (mA/cm2)	V_{OC} (V)	FF	PCE (%)	References
FTO/TiO_2/Dye/Cs_2SnI_6 + N719/Pt	14.7	0.63	0.68	6.32	Lee et al. (2014)
ITO/ZnO/$MASnI_3$/ spiro-OMeTAD/Au	11.1	0.97	–	7.66	Bansode et al. (2015)
ITO/PEDOT: PSS/ $MAPb0.85Sn_{0.15}I_{3-y}Cl_y$/ $PC_{61}BM$/Ag	19.1	0.76	0.66	9.8	Zuo et al. (2014)
FTO/TiO_2/N719 Dye/ Perovskite/ZnO	19.2	0.72	0.73	10.2	Chung et al. (2012)
FTO/PEDOT: PSS/ $EA_{0.98}EDA_{0.01}SnI_3$/C_{60}BCP/ Au	20.32	0.84	0.78	13.24	Mohanty et al. (2021)
ITO/PEDOT: PSS/C_{60}BCP/ Ag	26.1	0.74	0.71	14.1	Eperon et al. (2016)
ITO/PEDOT: PSS/ $MA_{0.4}FA_{0.6}Sn_{0.6}Pb_{0.4}I_3$/ C_{60}BCP/Ag	29.86	0.79	0.70	15.08	Liao et al. (2016)
ITO/PEDOT: PSS/ $MAPb0.85In_{0.15}I_{0.15}Cl_{0.15}$/ $PC_{61}BM$/Bphen/Ag	21.9	1.03	0.78	17.55	Wang et al. (2016b)
FTO/SnO_2/ $Cs_{0.16}FA_{0.84}Pb(I_{0.88}Br_{0.12})_3$/ spiro-OMeTAD/Au	22.4	1.02	0.78	18	Fievez et al. (2021)
FTO/Pply-TPD/Al-doped $MAPbI_3$/PCBM/BCP/Ag	22.4	1.01	0.78	19.1	Wang et al. (2016a)
FTO/*c*-TiO_2/*m*-TiO_2/$FAPbI_3$/ spiro-OMeTAD/Au	26.35	1.16	0.8	24.8	Jeong et al. (2020)
FTO/SnO_2-Cl/$FAPbI_3$/ spiro-OMeTAD/Au	25.74	1.18	0.83	25.5	Feng et al. (2021)

circuit connection (i.e., with zero current), and it is known as the V_{OC}. On the other hand, carriers flow through a short circuit (zero voltage) formed between the two semiconductor terminals, producing an electrical current known as the short-circuit current (I_{SC}) (Rahman et al. 2022).

An association between the voltage and current on the p–n junction is established by Eq. 1.9. It consists of two terms: one that represents the observed current when a voltage is applied, and the other that characterizes the behavior of the semiconductor under illumination ($I_{ph} \approx I_{SC}$). The reverse saturation current of the p–n junction also referred to as the dark current in photovoltaic applications is included in the

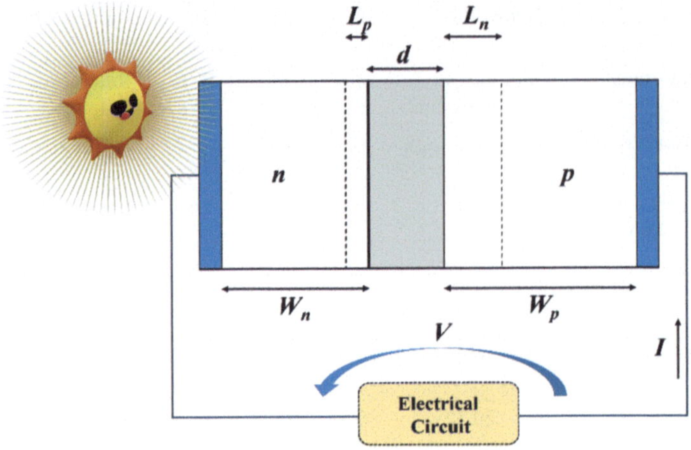

Fig. 1.15 Typical schematics of a p–n junction solar cell. Reproduced from Ref. Rahman et al. (2022), Copyright (2022) with permission of MDPI

latter term, I_{is}, and the junction non-ideality factor, represented by the symbol n. Additionally, Eq. 1.9 involves the thermal voltage (μ_{T}), described as Eq. 1.10 for a given temperature T, where K denotes Boltzmann's constant, and q denotes the electric charge.

$$I = I_{\mathrm{is}}e^{\frac{V}{n\mu_{\mathrm{T}}}} - 1 - I_{\mathrm{ph}} \tag{1.9}$$

$$\mu_{\mathrm{T}} = \frac{K}{q}T \tag{1.10}$$

Equation 1.9 applies only to the directions indicated by the arrows in Fig. 1.15, where the transition region is thought to be the place where all of the applied voltage V dissipates. This supposition stems from the idea that resistance on ohmic contacts and outside the transition area can be written off as insignificant. The value of I_{ph} is reliant on several factors, such as the illumination conditions, the device structure, and the type of semiconductors used. Assuming consistent monochromatic light, a constant photoelectric generation rate, G_{ph} of electron–hole pairs, and mild injection, a simple analysis can be performed. By analyzing the evolution of minority carrier densities with respect to distance in both the n and p regions and applying the diffusion equation for both carriers at the interfaces between the transition region and other regions, it is possible to calculate the short-circuit current. Equation 1.11 is applicable when $W_{\mathrm{p}} >> L_{\mathrm{n}}$ and $W_{\mathrm{n}} >> L_{\mathrm{p}}$, whereas Eq. 1.12 holds when $W_{\mathrm{p}} << L_{\mathrm{n}}$ and $W_{\mathrm{n}} << L_{\mathrm{p}}$.

$$I_{\mathrm{ph}} = qG_{\mathrm{ph}}L_{\mathrm{n}} + L_{\mathrm{p}}A \tag{1.11}$$

$$I_{\mathrm{ph}} = qG_{\mathrm{ph}}W_{\mathrm{n}} + W_{\mathrm{p}}\frac{A}{2} \tag{1.12}$$

In the equations above, the parameters L_{n} and L_{p} correspond to the diffusion lengths, while W_{n} and W_{p} represent the widths of the respective regions. A denotes the cross-sectional area of the device. The junction can only produce electricity when the voltage is positive and the current is negative, as indicated by the voltage and current directions in Fig. 1.15. Otherwise, the junction dissipates energy. Notably, the absence of illumination (i.e., when the photon current, $I_{\mathrm{ph}} = 0$) renders power production impossible. This is due to the fact that, for a positive voltage and a negative current, the current value is close to the saturation current (I_{is}), resulting in practically unattainable and extremely low power output.

$$I = I_{\mathrm{ph}} - I_{\mathrm{is}}e^{\frac{V}{n\mu_{\mathrm{T}}}} - 1 \tag{1.13}$$

It is possible to derive Eq. 1.13 by inverting the current signal from Eq. 1.9 in order to examine the junction as a solar cell. This modified equation (i.e., Eq. 1.13) is commonly employed in photovoltaic applications, wherein power is generated only when both V and I exhibit positive values. Furthermore, it is pertinent to note that the short-circuit current should be referred to as the photo-generated current (I_{ph}), as it may exist for various load resistance values.

1.3.5 Electric Models

The electrical model of a solar cell is essential for illustrating how a device functions under certain circumstances and with certain presumptions. When it comes to solar cells, the quantity of radiant flux received per unit area is known as the input variable (irradiance), and there are a number of internal factors in addition to the two output variables, V and I. The values of these internal parameters are subject to variations based on the specific model being employed and the phenomena being modeled.

The working principle outlined in Sect. 3.4 is used to construct the first electrical model, which entails the parallel conjunction of a diode and a distinctive current source, as seen in Fig. 1.16a. These electrical components are connected in parallel with the load. Since its value relies on incident radiation but remains constant under specific modeling parameters, a separate current source is classified as such. The p–n junction itself is modeled using the diode. As a result, the model shown in Fig. 1.16a is the most basic representation of a solar cell and has three parameters: I_{ph}, n, and I_{is}. These parameters can be subject to variations depending on external conditions such as temperature or irradiance, as will be further discussed below. This model is commonly called 1M3P or 1D3P.

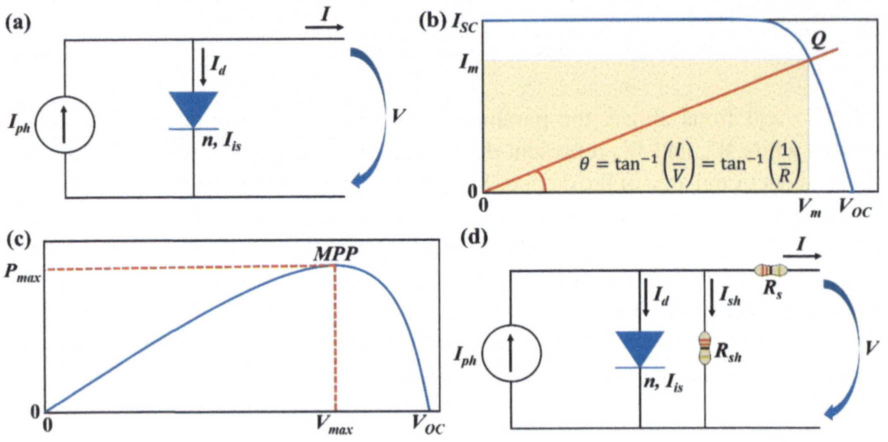

Fig. 1.16 **a** Schematically equivalent circuit of 1M3P model, **b**, **c** *I–V* and *P–V* characteristics correspond to 1M3P model, and **d** schematically equivalent circuit model of 1M5P model

Mathematically, Eq. 1.13 describes the relationship between a solar cell's output current and output voltage. This expression can be utilized as the basis for the electrical circuit. Conversely, the electrical circuit can be derived from the expression using Kirchhoff's voltage and current laws. The *I–V* curve obtained from Eq. 1.13 is shown in Fig. 1.16b, representing the relationship between the current and voltage that the solar cell generates. One resistance, *R*, can be used to signify the load circuit while the solar cell generates direct current (DC) power. The operating point *Q* of the solar cell is determined by the intersection of the characteristic curve and the load line. A blue line represents the characteristic curve and includes the short-circuit current density (J_{SC}) and V_{OC} of the solar cell. A particular V and I that satisfy Ohm's law for the load resistance determine the operating point Q.

Figure 1.16c illustrates a *P–V* curve, which represents the relationship between the output DC power and the required/produced voltage of a solar cell. The output power is determined by the product of the current and voltage ($P = I \times V$). The point on the curve where the solar cell produces the most power output is known as the maximum power point, or MPP. It should be remembered that a particular load resistance determines the output V and I, which in turn affect the produced power. Therefore, operating the solar cell at its MPP with the appropriate load resistance is necessary to achieve maximum power output.

Equation 1.13 can be used to determine the characteristic points of a solar cell. By setting the voltage to zero ($V = 0$) in Eq. 1.12, the photo-generated current can be approximated as the short-circuit current, represented by Eq. 1.14. Likewise, by setting the current to zero ($I = 0$) in Eq. 1.13, the open-circuit voltage can be obtained, as described by Eq. 1.15. Thus, these expressions enable the determination of short-circuit current and open-circuit voltage, which are key parameters for characterizing a solar cell.

$$I_{SC} \approx I_{ph} \tag{1.14}$$

$$V_{OC} = n\mu_T \ln \frac{I_{ph}}{I_{is}} + 1 \tag{1.15}$$

The maximum power point of a solar cell can be determined by finding the point on the *P*–*V* curve where the derivative is zero. Since there is only one maximum point, it can be determined using Eq. 1.16, derived from Eq. 1.13. Further, Eq. 1.17 is obtained directly from Eq. 1.16, enabling voltage calculation at the maximum power point. By substituting this voltage value into Eq. 1.13, the corresponding current value can be determined, as described by Eq. 1.18. Multiplying these voltage and current values yields the output DC power at the maximum power point. Therefore, the maximum power point of the solar cell can be entirely determined using these expressions.

$$\left(\frac{dP}{dV}\right)_{V=V_{max}} = 0 \rightarrow I_{ph} + \frac{V_{max}}{n\mu_T} + 1I_{is}e^{\frac{V_{max}}{n\mu_T}} + I_{is} = 0 \tag{1.16}$$

$$V_{max} = n\mu_T \ln \frac{\frac{I_{ph}}{I_{is}} + 1}{\frac{V_{max}}{n\mu_T} + 1} \tag{1.17}$$

$$I_{max} = I_{ph} - I_{is}e^{\frac{V_{max}}{n\mu_T} - 1} \tag{1.18}$$

In addition to the characteristic points and maximum power points, other parameters are essential for analyzing and comparing solar cells. PCE (η) and fill factor (FF) are important parameters used to characterize solar cell performance. Efficiency indicates the percentage of incident solar energy converted into electrical energy, while the fill factor measures how effectively the cell converts light into electricity. These parameters provide valuable information on the overall performance and effectiveness of the cell and are commonly used to compare different types of cells.

The ratio of incident flux power to output-produced power is the definition of a solar cell's efficiency. To ascertain the efficiency of the cell for a specific load resistance, the maximum power point must be computed, as previously said. The irradiance, which is the radiant flux acquired per unit area scaled by the active area of the PV cell, can be used to calculate the input power, which is not an electrical variable. Consequently, the quantity of power incidence on the cell is determined by the irradiance. These relationships are presented by Eq. 1.19, which expresses the fundamental relationship between the input and output power of a solar cell.

$$\eta = \frac{P_{max}}{P_{inc}} = \frac{I_{max}V_{max}}{G_{inc}A_{active}} \tag{1.19}$$

The FF, expressed in Eq. 1.20, is a crucial parameter for evaluating solar cell performance. Its definition is the ratio between the maximum power output to the product of the J_{SC} and V_{OC}. For an ideal p–n junction, where the *I–V* characteristic forms a rectangle, a fill factor of 100% would be achieved. However, in practice, other types of solar cells, such as multi-junction cells, tend to exhibit lower FF than single p–n junction cells. Thus, the FF serves as a useful metric for comparing the performance of different solar cell technologies.

$$\mathrm{FF} = \frac{P_{\max}}{I_{SC} V_{OC}} = \frac{I_{\max} V_{\max}}{I_{SC} V_{OC}} \tag{1.20}$$

Given an *I–V* characteristic like the one depicted in Fig. 1.16b, it is possible to estimate the values of the cell's efficiency and the photocurrent at zero bias (I_{is}) by applying Eqs. 1.21 and 1.22, respectively, which are directly derived from Eq. 1.13. To do so, one must select two distinct points, denoted as (V_1; I_1) and (V_2; I_2). The choice of these points is critical, as different pairs may yield varying approximations. Thus, selecting points with differing voltage and current values is advisable.

$$\eta = \frac{V_2 - V_1}{\mu_{\mathrm{T}} \ln \frac{I_2 - I_{\mathrm{ph}}}{I_1 - I_{\mathrm{ph}}}} \tag{1.21}$$

$$I_{\mathrm{is}} = \frac{I_{\mathrm{ph}} - I_1}{e^{\frac{V_1}{n\mu_{\mathrm{T}}}}} = \frac{I_{\mathrm{ph}} - I_2}{e^{\frac{V_2}{n\mu_{\mathrm{T}}}}} \tag{1.22}$$

A solar cell, usually a straightforward p–n junction solar cell, is represented by the 1M3P model, which is referred as the simplest model of the PV cell. Although it can be applied to other types of solar cell technology, its accuracy may decrease. Loss parameters can be incorporated to enhance the 1M3P model. The 1M5P model, shown in Fig. 1.16d, is an improved equivalent model that includes two additional resistances compared to the 1M3P model.

The 1M3P model, an essential equivalent solar cell model, can be improved by adding some loss parameters. Since p–n junctions have leakage currents, an additional shunt resistance is included in the 1M3P model to account for these losses. The power losses associated with the leakage current can be computed using the expression $P_{\mathrm{sh}} = R_{\mathrm{s}} I_{\mathrm{sh}}^2$, where R_{sh} is the shunt resistance and I_{sh} is the equivalent leakage current. In addition, this circuit adds a second resistance, R_{s}, in series to compensate for power losses at the junction and ohmic contacts. These power losses can be computed as $P_{\mathrm{s}} = R_{\mathrm{s}}\, I$, with I as the output current. Using these circuit laws, Eq. 1.23 is derived.

$$I = I_{SC} - I_{\mathrm{is}} e^{\frac{V - R_{\mathrm{s}} I}{n\mu_{\mathrm{T}}}} - 1 - \frac{V + R_{\mathrm{s}} I}{R_{\mathrm{sh}}} \tag{1.23}$$

The output power can be determined by calculating the product of the output voltage and current. Several simulations were conducted for various values of R_{s} and R_{sh}, resulting in *I–V* and *P–V* curves, which illustrate the relationship between the current, voltage, and power. In the 1M5P scenario, the maximum power point of a solar cell can be obtained using Eq. 1.24, which enables the determination of the corresponding current and voltage. Additionally, Eq. 1.25 is derived from Eq. 1.23 by imposing the condition given in Eq. 1.24. This process is similar to the methodology for obtaining the maximum power point in the 1M3P scenario.

$$\left(\frac{dP}{dV}\right)_{V=V_{max}} = 0 \rightarrow I_{max} + \frac{-\frac{(R_{sh}I_{SC}-V_{OC}+R_sI_{SC})e^{\frac{V_{max}+R_sI_{max}-V_{OC}}{n\mu_T}}}{n\mu_T R_{sh}} - \frac{1}{R_{sh}}}{1+\frac{R_s(R_{sh}I_{SC}-V_{OC}+R_sI_{SC})e^{\frac{V_{max}+R_sI_{max}-V_{OC}}{n\mu_T}}}{n\mu_T R_{sh}} + \frac{R_s}{R_{sh}}} V_{max} = 0 \tag{1.24}$$

$$I_{max} = I_{SC} - \frac{V_{max} + R_s I_{max} - I_{SC}}{R_{sh}} - I_{SC} - \frac{V_{OC} - R_s I_{SC}}{R_{sh}} e^{\frac{V_{max}+R_sI_{max}-V_{OC}}{n\mu_T}} \tag{1.25}$$

However, to gain a more comprehensive understanding of the impact of these resistances on the characteristics, it is necessary to derive Eqs. 1.26 and 1.27. This would provide a scientific basis for analyzing the behavior of the solar cell under different conditions. In a lossless model such as 1M3P, where R_{sh} is considered to be infinite, the slope declines as resistance rises, suggesting a decrease in FF. By computing the slope of the *I–V* curve near the short-circuit point, Eq. 1.26 is obtained, revealing that solar cells that have greater R_{sh} have a lower slope in the *I–V* curve. This finding is linked to other solar cell specifications. For instance, a higher R_{sh} implies a more imperfect junction and, therefore, a lower FF. As a result, the *I–V* curve deviates from the ideal rectangular shape. Equation 1.27 is used to deduce that the *I–V* slope near the open-circuit point is affected by R_s. The higher the slope, the lower the resistance. This observation is consistent with the lossless model, 1M3P, where R_s is assumed to be zero.

$$\left(\frac{dI}{dV}\right)_{I=I_{SC}} = -\frac{1}{R_{sh}} \tag{1.26}$$

$$\left(\frac{dI}{dV}\right)_{V=V_{OC}} = -\frac{1}{R_s} \tag{1.27}$$

Adding resistances can impact other measurable parameters, such as the FF. Equation 1.28 recalculates FF, with FF0 representing the original fill factor value obtained using expression.

$$FF = FF_0 1 - \frac{R_s I_{SC}}{V_{OC}} 1 - \frac{V_{OC}}{R_{sh} I_{SC}} \tag{1.28}$$

The models that were previously mentioned offer a way to describe the electrical characteristics of basic solar cells, like p–n or *p–i–n* cells. However, alternative models may be required for more complex solar cells to achieve accurate results. In some cases, models featuring multiple diodes must account for additional loss

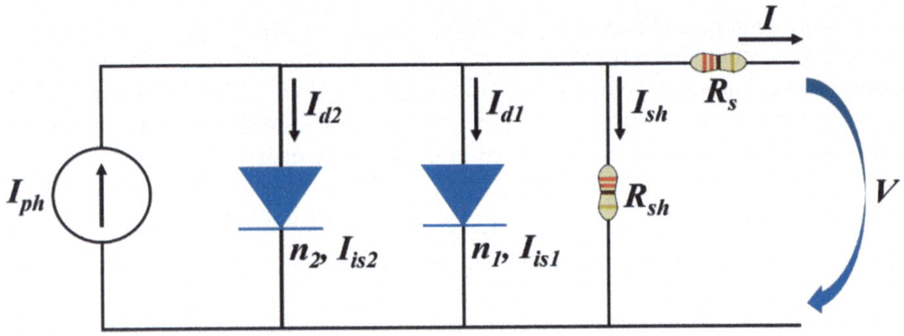

Fig. 1.17 Schematic electrical circuit of more rigorous or complex 1M7P or 2D7P model

phenomena. Moreover, numerous cell models may be combined to simulate a solar module or panel.

An example of a more complex model is the 1M7P or 2D7P model depicted in Fig. 1.17 (Rahman et al. 2022). This model considers not only losses due to recombination but also losses through diffusion, which can be simulated by subtracting a current from the I_{ph}. Therefore, an additional diode in parallel is included in the model. The output current expression for the photovoltaic cell illustrated in Fig. 1.17 is determined using a node analysis similar to the previous process. This results in the derivation of Eq. 1.29. Moreover, Eq. 1.28 can also be employed to compute the fill factor in this scenario. Since the added losses are included in the estimation of the initial fill factor value (FF_0, without resistance losses).

$$I = I_{\mathrm{ph}} - I_{d1} - I_{d2} - I_{\mathrm{sh}} = I_{\mathrm{ph}} - I_{\mathrm{is}_1} e^{\frac{V+R_{\mathrm{S}}I}{n_1 u_{\mathrm{T}}}} - 1 - I_{\mathrm{is}_2} e^{\frac{V+R_{\mathrm{S}}I}{n_2 u_{\mathrm{T}}}} - 1 - \frac{V + R_{\mathrm{s}}I}{R_{\mathrm{sh}}} \tag{1.29}$$

By applying short-circuit and open-circuit limits to Eq. 1.29, the corresponding expressions for I_{SC} and V_{OC} can be deduced. These expressions are given by Eqs. 1.30 and 1.31, respectively.

$$I_{\mathrm{SC}} = I_{\mathrm{ph}} - I_{\mathrm{is}_1} e^{\frac{R_{\mathrm{S}} I_{\mathrm{SC}}}{n_1 u_{\mathrm{T}}}} - 1 - I_{\mathrm{is}_2} e^{\frac{R_{\mathrm{S}} I_{\mathrm{SC}}}{n_2 u_{\mathrm{T}}}} - 1 - \frac{R_{\mathrm{s}} I_{\mathrm{SC}}}{R_{\mathrm{sh}}} \tag{1.30}$$

$$V_{\mathrm{OC}} = I_{\mathrm{ph}} - I_{\mathrm{is}_1} e^{\frac{V_{\mathrm{OC}}}{n_1 u_{\mathrm{T}}}} - 1 - I_{\mathrm{is}_2} e^{\frac{V_{\mathrm{OC}}}{n_2 u_{\mathrm{T}}}} - 1 R_{\mathrm{sh}} \tag{1.31}$$

1.3.6 Perovskite Film Fabrication

Especially in comparison with SSCs, PSCs use more readily available, less expensive materials and go through a more straightforward production chain. A strongly ventilated environment that must reach temperatures beyond 1000 °C is necessary to manufacture SSCs (Al et al. 2020). On the other hand, PSCs can be produced using straightforward wet chemistry without an evacuated atmosphere. The circumstances and methods used to make the perovskite layer are essential for crystallinity, which controls how well the cell works. The environmental variables throughout film formation, reagents elemental composition, a technique to lay perovskite on the surface, annealing temperature/duration, and additives utilized have been recognized as the variables that must be rigorously controlled. There are three very famous perovskite film fabrication techniques: vapor deposition, solution-processed deposition, and hybrid deposition (Srivastava et al. 2023a).

In all the reported fabrication methodologies, solution processing technology is the most popular and economical approach for perovskite film coatings. Such coatings can be attained either in one-step or two-step solution coating via spin coating. In a one-step layer, commonly the mixed solution of MAX and PbX_2 (1:1) was coated on the preheated ETL/HTL substrate, while in a two-step coating, only PbX_2 was coated on the substrate and then submerged into a MAX solution. The available literature confirms that one-step coating is one of the cheap and effective ways to coat a homogeneous perovskite film with excellent control over morphology and crystallinity (Burschka et al. 2013). However, two-step coated perovskite film produced better performance, and one-step coating is widely utilized to coat perovskite films. Apart from these one- and two-step spin-coating methods, there are several other fascinating solution processing techniques such as spray coating, ink-jet printing, slot-die coating, blade coating, knife-over edge coating, and vacuum flash-assisted coating as depicted in Fig. 1.18.

On the other hand, in vapor deposition, the planar perovskite films were prepared via simultaneous evaporation of MAX and PbX_2. Liu et al. demonstrated the $MAPbI_3$ perovskite film preparation by co-evaporating MAI and $PbCl_2$ at 120 and 350 °C, respectively, with high-quality crystalline and uniform films (Liu et al. 2013). However, perovskite film production via this method is not economically feasible and is the main challenge for large-scale production. Besides co-evaporation, vapor deposition includes vacuum thermal evaporation, sequential evaporation, flash evaporation, vapor-assisted solution process, and chemical vapor deposition (Fig. 1.19).

Recently, Smecca et al. demonstrated a novel two-step low vacuum proximity space effusion (LV-PSE) for high-quality film production at meager expenses. This phenomenon follows the mechanism of adsorption-incorporation-migration following the MAX concentration gradient (Smecca et al. 2021). Recently, highly efficient and reproducible triple cation PSCs were developed via an industry-compatible two-step blade coating technique based on green anti-solvent quenching. Ink-jet printing based on meniscus-assisted solution printing was demonstrated by Ming et al. to control the orientation and size, which attained a PCE of 20% by reducing the trap states and enhancing charge extraction (He et al. 2017). The sulfolane additive-based one-step solution processing method was explored for high-quality uniform perovskite film coating. Using this solution-processed method, the reported PCE was 17.58 and 16.06% for both mini- and large-modules (Huang et al. 2021). However, after that, several other additives were explored and documented.

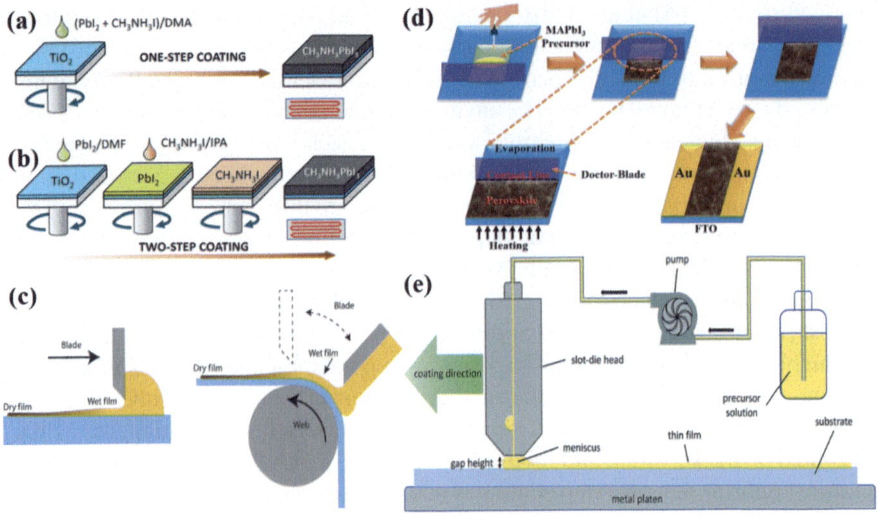

Fig. 1.18 Schematics of the various solution and vacuum processing-based PSC fabrication techniques. **a**, **b** One- and two-step spin-coating technique, Reproduced from Ref. Im et al. (2014), Copyright (2014) with permission of the American Institute of Physics. **c** Slot-die coating process, Reproduced from Ref. Swartwout et al. (2019), Copyright (2019) with permission of WILEY-VCH Verlag GmbH & Co. KGaA, Weinheim. **d** Doctor-blade coating, Reproduced from Ref. Srivastava et al. (2023b), copyright (2023) with permission of Elsevier. **e** Typical steps involved in the slot-die coating method, Reproduced from Ref. Swartwout et al. (2019), Copyright (2019) with permission of WILEY-VCH Verlag GmbH & Co. KGaA, Weinheim

Even spray and blade coatings were ancient technologies, but they still have utility. Ink-jet printing and slot-die coating methods are the most demanding and superior for large-area PSC fabrication. Slot-die-coated perovskite film produces the highest performance in large-area PSCs, while Si-perovskite tandem PSCs have produced ~30% PCE. In 2020, Ashouri and co-workers developed monolithic Si-perovskite PSCs with PCE > 29% and effective hole exportation (Al-Ashouri et al. 2020).

In conclusion, perovskite solar cells can be fabricated using solution-based or vapor-based techniques, each with its advantages and disadvantages. Solution-based techniques offer low-cost and high-throughput fabrication but may have limited reproducibility and stability issues. Vapor-based techniques provide better control over the film growth and can yield high-quality films, but they can be more expensive and time-consuming. As perovskite solar cell technology advances, a combination of these techniques may be utilized to achieve optimal performance and cost-effectiveness. One should follow review articles based on the perovskite fabrication processes to understand the depth of the various fabrication processes.

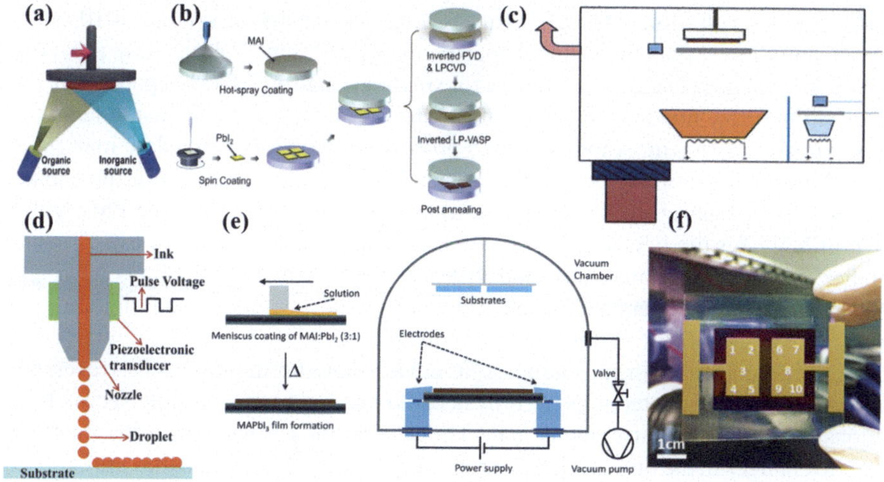

Fig. 1.19 Typical schematics of various vapor deposition processes. **a** CVD process, Reproduced from Ref. Chen et al. (2018), Copyright (2018) with permission of the Royal Society of Chemistry. **b** Vapor-solid reaction process, Reproduced from Ref. Chen et al. (2016), Copyright (2016) with permission of the Royal Society of Chemistry. **c** Vacuum thermal evaporation process, Reproduced from Ref. Wang et al. (2015), Copyright (2015) with permission of the Royal Society of Chemistry. **d** Ink-jet printing method, Reproduced from Ref. Karunakaran et al. (2019), Copyright (2019) with permission of the Royal Society of Chemistry. **e** Vacuum flash evaporation technique, Reproduced from Ref. Longo et al. (2015), Copyright (2015) with permission of the Royal Society of Chemistry. **f** 2.02 cm^2 PSC fabricated using an ink-jet printing technique, Reproduced from Ref. Li et al. (2018), Copyright (2018) with permission of Elsevier

1.4 Major Technological Advancement and Key Challenges

In recent years, metal halide perovskites have emerged as promising materials for photovoltaic technology. These materials are called hybrid perovskites because they are a combination of organic and inorganic compounds. The first hybrid perovskite-based solar cells were reported in 2009. The most commonly used hybrid perovskite materials as an absorber layer in PSC are $MAPbI_3$ and $MAPbBr_3$, which have bandgaps of 1.6 eV and 2.3 eV, respectively. In 2020, perovskite solar cells achieved a high PCE record of 25.5%, comparable to crystalline-silicon-based solar cells. Hybrid perovskite solar cells and semiconductors have recently gained tremendous attention as the fastest-growing photovoltaic technology. In addition to their use in solar cells, halide perovskites are suitable for optoelectronic devices such as light-emitting diodes (LEDs) and photodetectors.

Halide perovskite semiconductors possess several desirable properties that make them suitable for use in solar cells. These properties include strong absorption in the visible region, a long carrier diffusion length of up to approximately 1 μm, a weak exciton binding energy of around 45 meV, a high carrier mobility of roughly 25 $cm^2\ V^{-1}\ s^{-1}$, and a low charge recombination rate on the microsecond time scale. Single-crystalline $MAPbI_3$ exhibits a defect density of around 1010 cm^{-3}, significantly lower than the value of approximately 10^{16} cm^{-3} observed in polycrystalline perovskites. The carrier lifetime for single-crystalline $MAPbI_3$ perovskite can reach up to a few hundred microseconds, with a corresponding diffusion length of approximately 175 μm. Another unique property of $MAPbI_3$ perovskite material is its ambipolar transport property, which enables it to serve as a hole transport material. In 2012, Etgar et al. reported hole transport material (HTM) free PSC with a PCE of 5.5% for the first time. Many research groups have conducted experimental studies on perovskite materials, and different theoretical investigations have also been carried out to understand the electronic properties, potential, and limitations of perovskite materials for various device applications.

Despite their promising properties, PSC devices have limitations such as reversible current–voltage characteristics (*I*–*V* hysteresis) and long-term stability. These limitations arise due to anharmonicity in the lattice of halide perovskites. Anharmonicity leads to changes in the optoelectronic properties of perovskites when they are exposed to electric bias or light for an extended period. The resulting charge screening decreases the charge carrier recombination rate and weakens the B–X bonds, facilitating ion migration in mixed-halide perovskites, which leads to *I*–*V* hysteresis. Structural fluctuations of the lattice of halide perovskite are strongly coupled to their electronic states, and large polarons (about 9 nm) form around charge carriers due to the lattice distortions. These large polarons self-trap charge carriers, creating energetic barriers for charge carrier recombination and inhibiting electron and hole recombination because their polarization clouds screen the Coulomb potentials between the carriers. This screening effect causes long carrier lifetimes in halide perovskites. It may also protect charge carriers from being trapped at ionized defects, making them defect-tolerant despite being processed from a solution.

Halide PSCs have been used in tandem structures with other types of solar cells, such as silicon (Si) and copper indium gallium di-selenide (CIGS) solar cells, to achieve high efficiency. For example, a perovskite/silicon tandem solar cell was developed by Mailoa et al. (2015), with a stable efficiency of 13.7%. However, the main challenge for these tandem cells is achieving a perfect interface between the organic layer and the Si surface. The spin-coating method of depositing the perovskite film does not produce a uniform layer on textured crystalline silicon (c-Si) cells. Hence, Ballif et al. used the thermal evaporation method to achieve an efficiency of 25.2% with negligible hysteresis. However, the high efficiency of these cells decreased to around 22% after 60 h of testing due to degradation of the perovskite layer. One well-known issue with PSCs is their instability in humidity, which causes rapid degradation of the halide perovskite absorber layer. $MAPbI_3$ perovskite degrades in the presence of moisture and oxygen, producing various volatile products. The degradation mechanism of single-crystal $MAPbI_3$ perovskite has also been studied, with exposure to 80% relative humidity resulting in the transformation to metastable $MAPbI_3$–H_2O monohydrate. Eventually, excess water irreversibly degrades the perovskite structure. Sulfate treatment of perovskite film has been shown to effectively stabilize the film surface and bulk, improving water resistance. Another well-known issue is UV-induced degradation at the interface of TiO_2 and $MAPbI_3$ in mesoporous TiO_2-based PSCs, where TiO_2 extracts electrons from iodide and dissociates the perovskite structure.

The degradation of PSC is a critical issue that hinders their commercialization. One of the leading causes of degradation is the diffusion of metal electrodes (Ag or Au) into the absorber layer of the PSC. Moisture present in the environment can cause the halide perovskite ($MAPbI_3$) to degrade and release hydrogen iodide (HI) through the pinholes of the charge transport layer (e.g., PCBM). The HI can then react with the top Ag electrode to form silver iodide (AgI), which further diffuses into the solar cell and deteriorates its performance and stability. In the case of the Au electrode, the reactive polyiodide melts $[I_3^-]$ released during the degradation of iodide-based perovskite can react with the Au metal electrode to form $[AuI_2]^-$ and $[AuI_4]^-$ complexes, which can also worsen the performance of PSC. It is worth noting that the Ag electrode diffuses more quickly than the Au electrode. Another issue is the light-induced phase segregation in mixed-halide perovskites, which creates trap states and leads to lower open circuit voltage (Voc) than expected. This type of photoinduced instability can have severe implications for PSC operation. To address these problems, PSCs are generally fabricated in a glove box under controlled humidity and oxygen levels. Various factors, including perovskite composition, ambient temperature, and UV radiation in the atmosphere, influence the stability of PSC.

While moisture is known to cause degradation in PSC, some studies have shown that moderate moisture levels (around 50% relative humidity) during the annealing process can benefit the device without affecting its performance. However, the degree to which moisture can be tolerated varies depending on the perovskite composition and deposition method. Furthermore, the stability of PSC is strongly influenced by interfacial defects, selective contact and metal electrode stability, light-induced degradation, and instability under bias, all of which are unavoidable factors during device operation, even with encapsulation, to exclude humidity. These factors pose critical challenges to achieving long-term stability and performance of PSCs. For instance, while some studies have reported improved PCE and stability with the presence of moisture during solvent vapor annealing, the issues of metal electrode diffusion, light-induced phase segregation, and instability under bias continue to be major concerns that must be addressed to realize the commercial viability of PSC technology.

Achieving high performance in PSCs requires careful consideration of various factors during fabrication. Despite their promising efficiency, the toxic element Pb in halide perovskites has raised environmental concerns, necessitating the search for alternative, non-toxic materials. Tin (Sn) has been proposed as a potential substitute for Pb, and other approaches involving polymer-based materials and nanomaterials have also been explored to improve efficiency and stability. Materials such as PCBM and PEDOT: PSS have been identified as advantageous due to their low-temperature processing requirements. Researchers have investigated PSC stability under both inert and ambient conditions. However, long-term stability, hysteresis effects, and large-scale production remain significant challenges to be addressed for the commercialization of PSCs.

1.5 Conclusion

This chapter examines photovoltaic technology, beginning with a historical overview, which reveals that photovoltaics were only sometimes a viable solution for energy needs and sustainability concerns. But as time went on, the technology improved and became a highly respected area of research since, in comparison with other renewable energy sources, it was comparatively inexpensive to fabricate, install, and maintain. Moreover, several distinct areas of photovoltaic technology exist, from space applications to public illumination, everyday devices, and large-scale photovoltaic parks.

The analysis goes over how a p–n junction functions as a solar cell and goes over a number of solar cell models, such as 1M3P and 1M5P and more sophisticated models like 1M7P. The article looks at how solar cell performance might change under unfavorable circumstances such as temperature and irradiance changes, cell breakdown, or shadowing because real-world conditions are rarely perfect or static. In order to find out how these phenomena affect solar cells experimentally, a literature review is done and these effects are quantitatively examined using electrical models.

To attract fresh researchers to this fascinating topic, this chapter aims to inform readers about significant turning points in the history of photovoltaics, summarize its foundations, and highlight some astonishing applications.

References

Abbel R, Galagan Y, Groen P (2018) Roll-to-roll fabrication of solution processed electronics. Adv Eng Mater 20:1–30. https://doi.org/10.1002/adem.201701190

Al-Ashouri A, Köhnen E, Li B et al (2020) Monolithic perovskite/silicon tandem solar cell with > 29% efficiency by enhanced hole extraction. Science (80-) 370:1300–1309. https://doi.org/10.1126/science.abd4016

Al MG et al (2020) Recent progress towards roll-to-roll manufacturing of perovskite solar cells using slot-die processing. Flex Print Electron 5:014006. https://doi.org/10.1088/2058-8585/ab639e

Alves dos Santos SA, João JP, Carlos CA, Marques Lameirinhas RA (2021) The impact of aging of solar cells on the performance of photovoltaic panels. Energy Convers Manag X 10. https://doi.org/10.1016/j.ecmx.2021.100082

Andreev VM et al (2003) GaAs and high-efficiency space cells. In: Practical handbook of photovoltaics, pp 417–433. https://doi.org/10.1016/B978-185617390-2/50017-9

Bansode U, Naphade R, Game O et al (2015) Hybrid perovskite films by a new variant of pulsed excimer laser deposition: a room temperature dry process. J Phys Chem C 119:9177–9185. https://doi.org/10.1021/acs.jpcc.5b02561

Benda V, Černá L et al (2020) PV cells and modules—state of the art, limits and trends. Heliyon 6. https://doi.org/10.1016/j.heliyon.2020.e05666

Bernardes S, Marques Lameirinhas RA, Torres JPN, Fernandes CAF (2021) Characterization and design of photovoltaic solar cells that absorb ultraviolet, visible and infrared light. Nanomaterials 11:1–16. https://doi.org/10.3390/nano11010078

Burschka J, Pellet N, Moon SJ et al (2013) Sequential deposition as a route to high-performance perovskite-sensitized solar cells. Nature 499:316–319. https://doi.org/10.1038/nature12340

Burst JM, Duenow JN, Albin DS et al (2016) CdTe solar cells with open-circuit voltage breaking the 1V barrier. 1. https://doi.org/10.1038/NENERGY.2016.15

Cao J, Yan F et al (2021) Recent progress in tin-based perovskite solar cells. Energy Environ Sci 14:1286–1325. https://doi.org/10.1039/d0ee04007j

Carmo JE, Torres JPN, Cruz G, Marques Lameirinhas RA (2021) Effect of the inclusion of photovoltaic solar panels in the autonomy of UAV time of flight. Energies 14:1–19. https://doi.org/10.3390/en14040876

Chand Yadav S, Srivastava A, Manjunath V et al (2022) Properties, performance and multidimensional applications of stable lead-free $Cs_2AgBiBr6$ double perovskite. Mater Today Phys 26:100731. https://doi.org/10.1016/j.mtphys.2022.100731

Chen M, Shao L, Kershaw SV et al (2014) Photocurrent enhancement of HgTe quantum dot photodiodes by plasmonic gold nanorod structures. ACS Nano 8:8208–8216. https://doi.org/10.1021/nn502510u

Chen Q, De Marco N, Yang Y et al (2015) Under the spotlight: the organic-inorganic hybrid halide perovskite for optoelectronic applications. Nano Today 10:355–396. https://doi.org/10.1016/j.nantod.2015.04.009

Chen X, Cao H, Yu H et al (2016) Large-area, high-quality organic-inorganic hybrid perovskite thin films: via a controlled vapor-solid reaction. J Mater Chem A 4:9124–9132. https://doi.org/10.1039/c6ta03180c

Chen Y, Zhang L, Zhang Y et al (2018) Large-area perovskite solar cells-a review of recent progress and issues. RSC Adv 8:10489–10508. https://doi.org/10.1039/c8ra00384j

Chopra KL, Paulson PD, Dutta V (2004) Thin-film solar cells: an overview. Prog Photovoltaics Res Appl 12:69–92. https://doi.org/10.1002/pip.541

Chung I, Lee B, He J et al (2012) All-solid-state dye-sensitized solar cells with high efficiency. Nature 485:486–489. https://doi.org/10.1038/nature11067

da Silva Ravasco A, Torres JPN, Lameirinhas R (2021) Wearable photovoltaic applications as energy sources for everyday devices. Am J Eng Appl Sci 14:337–350. https://doi.org/10.3844/ajeassp.2021.337.350

Das KK, Reddy RC, Bagoji IB et al (2019) Primary concept of nickel toxicity—an overview. J Basic Clin Physiol Pharmacol 30:141–152. https://doi.org/10.1515/jbcpp-2017-0171

Efros AL, Rosen M et al (1998) Quantum size level structure of narrow-gap semiconductor nanocrystals: effect of band coupling. Phys Rev B—Condens Matter Mater Phys 58:7120–7135. https://doi.org/10.1103/PhysRevB.58.7120

Einstine A (1920) NobelPrize.org Nobel prize outreach AB

Elseman AM, Sajid S, Shalan AE et al (2019) Recent progress concerning inorganic hole transport layers for efficient perovskite solar cells. Appl Phys A Mater Sci Process 125:1–12. https://doi.org/10.1007/s00339-019-2766-7

Eperon GE, Eperon GE, Leijtens T et al (2016) Perovskite-perovskite tandem photovoltaics with optimized bandgaps. Science (80-) 9717:1–10

Evans HA, Wu Y, Seshadri R, Cheetham AK (2020) Perovskite-related ReO_3-type structures. Nat Rev Mater 5:196–213. https://doi.org/10.1038/s41578-019-0160-x

Feng X, Guo Q, Xiu J et al (2021) Close-loop recycling of perovskite solar cells through dissolution-recrystallization of perovskite by butylamine. Cell Reports Phys Sci 2:100341. https://doi.org/10.1016/j.xcrp.2021.100341

Feurer T, Reinhard P, Avancini E et al (2017) Progress in thin film CIGS photovoltaics—research and development, manufacturing, and applications. Prog Photovoltaics Res Appl 25:645–667. https://doi.org/10.1002/pip.2811

Fievez M, Singh Rana PJ, Koh TM et al (2021) Slot-die coated methylammonium-free perovskite solar cells with 18% efficiency. Sol Energy Mater Sol Cells 230. https://doi.org/10.1016/j.solmat.2021.111189

Fraunhofer ISE (2021) Photovoltaics report, Photovoltaics Report - Fraunhofer ISE

Ganesan AA, Houtepen AJ, Crisp RW (2018) Quantum dot solar cells: small beginnings have large impacts. Appl Sci 8. https://doi.org/10.3390/app8101867

Gao J, Fidler AF, Klimov VI (2015) Carrier multiplication detected through transient photocurrent in device-grade films of lead selenide quantum dots. Nat Commun 6:1–8. https://doi.org/10.1038/ncomms9185

Geisz JF, France RM, Schulte KL et al (2020) Six-junction III–V solar cells with 47.1% conversion efficiency under 143 Suns concentration. Nat Energy 5:326–335. https://doi.org/10.1038/s41560-020-0598-5

Gerischer H et al (1990) The impact of semiconductors on the concepts of electrochemistry. Electrochim Acta 35:1677–1699. https://doi.org/10.1016/0013-4686(90)87067-C

Ghosh A et al (2020) Potential of building integrated and attached/applied photovoltaic (BIPV/BAPV) for adaptive less energy-hungry building's skin: a comprehensive review. J Clean Prod 276:123343. https://doi.org/10.1016/j.jclepro.2020.123343

Giansante C, Infante I, Fabiano E et al (2015) "darker-than-black" PbS quantum dots: enhancing optical absorption of colloidal semiconductor nanocrystals via short conjugated ligands. J Am Chem Soc 137:1875–1886. https://doi.org/10.1021/ja510739q

Gordillo G et al (1992) Photoluminescence and photoconductivity studies on $Zn_xCd_{1-x}S$ thin films. Sol Energy Mater Sol Cells 25:41–49. https://doi.org/10.1016/0927-0248(92)90015-H

Green MA, Dunlop ED, Hohl-Ebinger J et al (2021) Solar cell efficiency tables (Version 58). Prog Photovoltaics Res Appl 29:657–667. https://doi.org/10.1002/pip.3444

He M, Li B, Cui X et al (2017) Meniscus-assisted solution printing of large-grained perovskite films for high-efficiency solar cells. Nat Commun 8. https://doi.org/10.1038/ncomms16045

Hens Z, Moreels I et al (2012) Light absorption by colloidal semiconductor quantum dots. J Mater Chem 22:10406–10415. https://doi.org/10.1039/c2jm30760j

Hiramoto M, Fujiwara H, Yokoyama M (1991) Three-layered organic solar cell with a photoactive interlayer of codeposited pigments. Appl Phys Lett 58:1062–1064. https://doi.org/10.1063/1.104423

Hoefler SF, Trimmel G, Rath T (2017) Progress on lead-free metal halide perovskites for photovoltaic applications: a review. Monatshefte fur Chemie 148:795–826. https://doi.org/10.1007/s00706-017-1933-9

Hoyer P, Könenkamp R (1995) Photoconduction in porous TiO_2 sensitized by PbS quantum dots. Appl Phys Lett 351:349. https://doi.org/10.1063/1.114209

Huang TJ, Thiang ZX, Yin X et al (2016) $(CH_3NH_3)_2PdCl_4$: a compound with two-dimensional organic-inorganic layered perovskite structure. Chem—A Eur J 22:2146–2152. https://doi.org/10.1002/chem.201503680

Huang HH, Liu QH, Tsai H et al (2021) A simple one-step method with wide processing window for high-quality perovskite mini-module fabrication. Joule 5:958–974. https://doi.org/10.1016/j.joule.2021.02.012

IEA (2010) Technology roadmap–solar photovoltaic energy

Im JH, Kim HS, Park NG (2014) Morphology-photovoltaic property correlation in perovskite solar cells: one-step versus two-step deposition of $CH_3NH_3PbI_3$. APL Mater 2. https://doi.org/10.1063/1.4891275

IRENA (2022) IRENA. Renew Energy Technol, IRENA – International Renewable Energy Agency

Jeong M, Choi IW, Go EM et al (2020) Stable perovskite solar cells with efficiency exceeding 24.8% and 0.3-V voltage loss. Science (80-) 369:1615–1620. https://doi.org/10.1126/science.abb7167

Ju T, Graham RL, Zhai G et al (2010) High efficiency mesoporous titanium oxide PbS quantum dot solar cells at low temperature. Appl Phys Lett 97. https://doi.org/10.1063/1.3459146

Kabir E, Kumar P, Kumar S et al (2018) Solar energy: potential and future prospects. Renew Sustain Energy Rev 82:894–900. https://doi.org/10.1016/j.rser.2017.09.094

Karunakaran SK, Arumugam GM, Yang W et al (2019) Recent progress in inkjet-printed solar cells. J Mater Chem A 7:13873–13902. https://doi.org/10.1039/c9ta03155c

Kieslich G, Sun S, Cheetham AK (2014) Solid-state principles applied to organic–inorganic perovskites: new tricks for an old dog. Chem Sci 5:4712–4715. https://doi.org/10.1039/c4sc02211d

Kim MS, Yun JH, Yoon KH, Ahn BT (2007) Fabrication of flexible CIGS solar cell on stainless steel substrate by co-evaporation process. Solid State Phenom 124–126:73–76. https://doi.org/10.4028/www.scientific.net/ssp.124-126.73

Lee B, Stoumpos CC, Zhou N et al (2014) Air-stable molecular semiconducting iodosalts for solar cell applications: Cs_2SnI_6 as a hole conductor. J Am Chem Soc 136:15379–15385. https://doi.org/10.1021/ja508464w

Li C, Lu X, Ding W et al (2008) Formability of ABX_3 (X = F, Cl, Br, I) halide perovskites. Acta Crystallogr Sect B Struct Sci 64:702–707. https://doi.org/10.1107/S0108768108032734

Li P, Liang C, Bao B et al (2018) Inkjet manipulated homogeneous large size perovskite grains for efficient and large-area perovskite solar cells. Nano Energy 46:203–211. https://doi.org/10.1016/j.nanoen.2018.01.049

Liao W, Zhao D, Yu Y et al (2016) Fabrication of efficient low-bandgap perovskite solar cells by combining formamidinium tin iodide with methylammonium lead iodide. J Am Chem Soc 138:12360–12363. https://doi.org/10.1021/jacs.6b08337

Liu M, Johnston MB, Snaith HJ (2013) Efficient planar heterojunction perovskite solar cells by vapour deposition. Nature 501:395–398. https://doi.org/10.1038/nature12509

Liu Q, Jiang Y, Jin K et al (2020a) 18% efficiency organic solar cells. Sci Bull 65:272–275. https://doi.org/10.1016/j.scib.2020.01.001

Liu Y, Li Y, Wu Y et al (2020b) High-efficiency silicon heterojunction solar cells: materials, devices and applications. Mater Sci Eng R Reports 142:1–116. https://doi.org/10.1016/j.mser.2020.100579

Liu Z, Liu P, Li M et al (2022) Efficient and stable FA-rich perovskite photovoltaics: from material properties to device optimization. Adv Energy Mater 12:1–35. https://doi.org/10.1002/aenm.202200111

Longo G, Gil-Escrig L, Degen MJ et al (2015) Perovskite solar cells prepared by flash evaporation. Chem Commun 51:7376–7378. https://doi.org/10.1039/c5cc01103e

Luther JM, Jain PK, Ewers T, Alivisatos AP (2011) Localized surface plasmon resonances arising from free carriers in doped quantum dots. Nat Mater 10:361–366. https://doi.org/10.1038/nmat3004

Mailoa et al (2015) A 2-terminal perovskite/silicon multijunction solar cell enabled by a silicon tunnel junction. Appl Phys Lett 106:121105. https://doi.org/10.1063/1.4914179

Melo I, Paulo J, Torres N et al (2020) Sustainability economic study of the islands of the Azores archipelago using photovoltaic panels, wind energy and storage system. Renew Wind Water Sol. https://doi.org/10.1186/s40807-020-00061-8

Metzger WK, Grover S, Lu D et al (2019) Exceeding 20% efficiency with in situ group V doping in polycrystalline CdTe solar cells. Nat Energy 4:837–845. https://doi.org/10.1038/s41560-019-0446-7

Michaels H, Benesperi I, Freitag M (2021) Challenges and prospects of ambient hybrid solar cell applications. Chem Sci 12:5002–5015. https://doi.org/10.1039/d0sc06477g

Miyasaka et al. (2022) Research background and recent progress of perovskite photovoltaics. Perovskite Photovoltaics Optoelectron Fundam Adv Appl (2022):1–60. https://doi.org/10.1002/9783527826391.ch1

Miyata A, Mitioglu A, Plochocka P et al (2015) Direct measurement of the exciton binding energy and effective masses for charge carriers in organic-inorganic tri-halide perovskites. Nat Phys 11:582–587. https://doi.org/10.1038/nphys3357

Mohanty I, Mangal S, Singh UP (2021) Performance optimization of lead free-$MASnI_3$/CIGS heterojunction solar cell with 28.7% efficiency: a numerical approach. Opt Mater (Amst) 122:111812. https://doi.org/10.1016/j.optmat.2021.111812

Nakamura M, Yamaguchi K, Kimoto Y et al (2019) Cd-Free Cu(In, Ga)(Se, $S)_2$ thin-film solar cell with record efficiency of 23.35 %. IEEE 9:1863–1867

Nick84 Wikimedia Commons (2023) Spectrum of solar radiation (Earth)

Nokia Bell Labs (1954) Bell Labs' greatest innovations

NREL. Best research-cell efficiency chart l Photovoltaic research l NREL

NREL (2023) Best research-cell efficiency chart

Ohl RS (1946) Light-sensitive electric device

O'Regan B, Grätzel M et al (1991) A low-cost, high-efficiency solar cell based on dye-sensitized colloidal TiO_2 films. Nature 354:737–740

Park NG et al (2018) Halide perovskite photovoltaics: history, progress, and perspectives. MRS Bull 43:527–533. https://doi.org/10.1557/mrs.2018.152

Pattantyus-Abraham AG, Kramer IJ, Barkhouse AR et al (2010) Depleted-heterojunction colloidal quantum dot solar cells. ACS Nano 4:3374–3380. https://doi.org/10.1021/nn100335g

Philipps S, Ise F et al 2023) PSE projects GmbH. Photovoltaics report

Qian J, Xu B, Tian W (2016) A comprehensive theoretical study of halide perovskites ABX_3. Org Electron 37:61–73. https://doi.org/10.1016/j.orgel.2016.05.046

Rahman UUH, Ghani U, Usman M, Ullah H, Khan A, El-Metwaly NM, Khan A (2022) A photovoltaic technology review: history, fundamentals and applications. Energies 15:1823. https://doi.org/10.3390/en15051823

Raj A, Kumar M et al (2021) Recent advancement in inorganic-organic electron transport layers in perovskite solar cell: current status and future outlook. Mater today Chem 22. https://doi.org/10.1016/j.mtchem.2021.100595

Reddy P, Gupta MVNS, Nundy S, Karthick A (2020) Status of BIPV and BAPV system for less energy-hungry building in India—a review. Appl Sci 10:2337

Riede M, Spoltore D, Leo K (2021) Organic solar cells—the path to commercial success. Adv Energy Mater 11:1–10. https://doi.org/10.1002/aenm.202002653

Roy P, Kumar Sinha N, Tiwari S, Khare A (2020) A review on perovskite solar cells: evolution of architecture, fabrication techniques, commercialization issues and status. Sol Energy 198:665–688. https://doi.org/10.1016/j.solener.2020.01.080

Roy P, Raoui Y, Khare A (2022) Design and simulation of efficient tin based perovskite solar cells through optimization of selective layers: theoretical insights. Opt Mater (Amst) 125:112057. https://doi.org/10.1016/j.optmat.2022.112057

Rühle S et al (2016) Tabulated values of the Shockley-Queisser limit for single junction solar cells. Sol Energy 130:139–147. https://doi.org/10.1016/j.solener.2016.02.015

Saparov B, Mitzi DB (2016) Organic-inorganic perovskites: structural versatility for functional materials design. Chem Rev 116:4558–4596. https://doi.org/10.1021/acs.chemrev.5b00715

Shannon RD (1976) Revised effective ionic radii and systematic studies of interatomic distances in halides and chalcogenides. Acta Crystallogr Sect B Struct Sci A32:751–767. https://doi.org/10.1107/S0567739476001551

Sharma S, Jain KK, Sharma A (2015) Solar cells: in research and applications—a review. Mater Sci Appl 06:1145–1155. https://doi.org/10.4236/msa.2015.612113

Shockley W, Queisser H (2018) Detailed balance limit of efficiency of p-n junction solar cells. Renew Energy Four Vol Set 2–4:35–54. https://doi.org/10.4324/9781315793245-44

Smecca E, Valenzano V, Valastro S et al (2021) Two-step $MAPbI_3$ deposition by low-vacuum proximity-space-effusion for high-efficiency inverted semitransparent perovskite solar cells. J Mater Chem A 9:16456–16469. https://doi.org/10.1039/d1ta02535j

Snaith HJ et al (2013) Perovskites: the emergence of a new era for low-cost, high-efficiency solar cells. J Phys Chem Lett 4:3623–3630. https://doi.org/10.1021/jz4020162

Srivastava A, Singh Chauhan B, Chand Yadav S et al (2022) Performance of dye-sensitized solar cells by utilizing Codiaeum Variegatum leaf and Delonix Regia flower as natural sensitizers. Chem Phys Lett 807:140087. https://doi.org/10.1016/j.cplett.2022.140087

Srivastava A, Kumar A, Tiwari K (2023a) Effect of $Ti_{1-x}Fe_xO_2$ photoanodes on the performance of dye-sensitized solar cells utilizing natural betalain pigments extracted from Beta vulgaris (BV). Energy Adv 2:148–160. https://doi.org/10.1039/d2ya00197g

Srivastava A, Satrughna JAK, Tiwari MK et al (2023b) Lead metal halide perovskite solar cells: Fabrication, advancement strategies, alternatives, and future perspectives. Mater Today Commun 35. https://doi.org/10.1016/j.mtcomm.2023.105686

Sun Q, Yin WJ et al (2017) Thermodynamic stability trend of cubic perovskites. J Am Chem Soc 139:14905–14908. https://doi.org/10.1021/jacs.7b09379

Swartwout R, Hoerantner MT, Bulović V (2019) Scalable deposition methods for large-area production of perovskite thin films. Energy Environ Mater 2:119–145. https://doi.org/10.1002/eem2.12043

Tang CW et al (1986) Two-layer organic photovoltaic cell. Appl Phys Lett 48:183–185. https://doi.org/10.1063/1.96937

Travis W, Glover ENK, Bronstein H et al (2016) On the application of the tolerance factor to inorganic and hybrid halide perovskites: a revised system. Chem Sci 7:4548–4556. https://doi.org/10.1039/c5sc04845a

Uribe JI, Ramirez D, Osorio-Guillén JM et al (2016) $CH_3NH_3CaI_3$ perovskite: synthesis, characterization, and first-principles studies. J Phys Chem C 120:16393–16398. https://doi.org/10.1021/acs.jpcc.6b04207

Vidyasagar CC, Muñoz Flores BM, Jiménez Pérez VM (2018) Recent advances in synthesis and properties of hybrid halide perovskites for photovoltaics

Wang S, Ono LK, Leyden MR et al (2015) Smooth perovskite thin films and efficient perovskite solar cells prepared by the hybrid deposition method. J Mater Chem A 3:14631–14641. https://doi.org/10.1039/c5ta03593g

Wang JTW, Wang Z, Pathak S et al (2016a) Efficient perovskite solar cells by metal ion doping. Energy Environ Sci 9:2892–2901. https://doi.org/10.1039/c6ee01969b

Wang ZK, Li M, Yang YG et al (2016b) High efficiency Pb–In binary metal perovskite solar cells. Adv Mater 28:6695–6703. https://doi.org/10.1002/adma.201600626

Wang Y, Wu T, Chueh Y (2019) A critical review on flexible Cu(In,Ga)Se_2 (CIGS) solar cells. 234:329–344

Wang R, Huang T, Xue J et al (2021) Prospects for metal halide perovskite-based tandem solar cells. Nat Photonics 15:411–425. https://doi.org/10.1038/s41566-021-00809-8

Werner J, Niesen B, Ballif C (2018) Perovskite/silicon tandem solar cells: marriage of convenience or true love story? An overview. Adv Mater Interfaces 5:1–19. https://doi.org/10.1002/admi.201700731

Wu T, Liu X, Luo X et al (2021) Lead-free tin perovskite solar cells. Joule 5:863–886. https://doi.org/10.1016/j.joule.2021.03.001

Yoshikawa K, Kawasaki H, Yoshida W et al (2017) Silicon heterojunction solar cell with interdigitated back contacts for a photoconversion efficiency over 26%. Nat Energy 2. https://doi.org/10.1038/nenergy.2017.32

Yu J, Wang W, Pan Z et al (2017) Quantum dot sensitized solar cells with efficiency over 12% based on tetraethyl orthosilicate additive in polysulfide electrolyte. J Mater Chem A 5:14124–14133. https://doi.org/10.1039/c7ta04344a

Zhang T, Wang M, Yang H (2018) A review of the energy performance and life-cycle assessment of building-integrated photovoltaic (BIPV) systems. Energies 11. https://doi.org/10.3390/en11113157

Zuo F, Williams ST, Liang PW et al (2014) Binary-metal perovskites toward high-performance planar-heterojunction hybrid solar cells. Adv Mater 26:6454–6460. https://doi.org/10.1002/adma.201401641

Chapter 2
Overcoming Intrinsic and Extrinsic Challenges in Perovskite Solar Cells: A Pathway Toward Advancement

Abstract This study underscores the urgency of tapping into Earth's solar abundance for electricity through renewable sources. As global solar PV capacity rises, the focus sharpens on boosting solar cell efficiency and cutting manufacturing costs. Third-generation Perovskite Solar Cells (PSCs) show promise to dominate with an anticipated 25.8% power conversion efficiency (PCE) in under a decade. Despite record efficiencies, the long-term stability of lead halide perovskites faces challenges from fragile bonds and mobile ions. Advanced techniques reveal insights into moisture, heat, and light-induced alterations on perovskite lattices. Efforts to systematically address degradation pathways are underway, with the potential for commercially competitive photovoltaic technology as multiple PSCs demonstrate strong performance and extended lifetimes.

Keywords Intrinsic stability: ion migration · Metal-perovskite interaction · Residual strain · Extrinsic stability: lead leakage · Humidity induced degradation · Thermal runaway

2.1 Introduction

PSCs have gained significant attention in the photovoltaic (PV) field due to their remarkable advancements in PCE. Despite this progress, silicon (Si) technology remains dominant, constituting approximately 95% of the global PV market. The primary challenge hindering the commercialization of PSCs lies in their limited operational stability, primarily attributed to intrinsic degradation mechanisms within perovskite materials. These degradation processes are associated with ion migration, metal-perovskite reactions, and residual strain in the perovskite lattices (Park 2015). As exemplified by $MAPbI_3$, organometal halide perovskites have emerged as a prominent class of PV materials for developing efficient, cost-effective, and solution-processed solar cells. These PSCs have garnered attention due to their exceptional electro-optical properties, such as high light absorption coefficients, extended carrier diffusion lengths, and superior carrier mobilities. This has facilitated significant

A. Srivastava and P. M. Shirage, *Perovskite Solar Cells*, Springer Tracts in Electrical and Electronics Engineering, https://doi.org/10.1007/978-981-97-7624-5_2

advancements in PSC performance, with PCEs soaring from 3.8% in 2009 to 25.8% in 2022, surpassing those achieved by conventional polycrystalline Si solar cells (NREL 2023). Despite this progress, the global PV market remains predominantly dominated by Si PV technology.

In evaluating the technical feasibility of PV technologies for commercialization, a comprehensive assessment encompassing cost, device efficiency, and device lifetime (stability) is crucial, as depicted in Fig. 2.1a (Meng et al. 2018). Figure 2.1b illustrates a comparison of device efficiencies for various PV technologies as a function of device area. Although the record efficiency of PSCs rivals that of crystalline Si solar cells and surpasses thin-film alternatives such as copper indium gallium selenide (CIGS) and cadmium telluride (CdTe), it is noteworthy that the champion PSCs possess a relatively small device area of less than 0.1 cm^2. In contrast, Si counterparts exhibit substantially larger device areas exceeding 100 cm^2. Furthermore, when considering PV modules, the record module efficiencies for Si (> 10,000 cm^2), CIGS (> 10,000 cm^2), and CdTe (> 10,000 cm^2) stand at 24.4%, 18.6%, and 19.0%, respectively (Werner et al. 2020). Moreover, a recent comprehensive analysis of current PV technologies has forecasted substantial growth in the market share of emerging PV technologies. Projections indicate that the emerging PV technologies are poised to capture a significant percentage of 44.1% by 2030, a substantial increase from a mere 1% in 2014, as illustrated in Fig. 2.1c. Within the realm of emerging PV technologies, PSCs have emerged as the most promising contender, showcasing their impressive device efficiencies that rival those of Si PV technology. However, it is crucial to note that the aforementioned market share prediction hinges upon the assumption that emerging PV panels exhibit a long-term operational lifespan of 25 years. Thus, to realize this projection, research endeavors in perovskite PV must shift their focus from solely advancing device efficiency to concurrently enhancing the device's lifetime (Chowdhury et al. 2020).

To comprehensively understand the inherent stability of organic–inorganic hybrid perovskites, it is crucial to gain profound insights into their lattice structure and the chemical bonding between the perovskite constituents. Figure 2.1d illustrates the lattice structure of ABX_3 perovskites, where the A-site can accommodate a monovalent organic or inorganic cation such as methylammonium (MA^+), formamidinium (FA^+), and Cs^+, the B-site can accommodate a divalent metal ion such as Pb^{2+} and Sn^{2+}, and the X-site can accommodate halide anions such as I^-, Br^-, and Cl^-. This arrangement establishes the foundation for the material's properties and stability. However, it should be noted that organic–inorganic hybrid perovskites are not inherently chemically robust due to the presence of weak ionic bonds and van der Waals interactions between the organic and inorganic components within the lattice structure. Consequently, these materials tend to decompose when subjected to various external factors and stimuli, as outlined in Fig. 2.1e. This degradation can occur under both environmental conditions and during device operation (Cheng et al. 2021).

While advanced encapsulation techniques can mitigate the detrimental effects of moisture and oxygen on the stability of PSCs, this chapter primarily focuses on the intrinsic stability of PSCs, explicitly addressing the chemical stability of perovskite

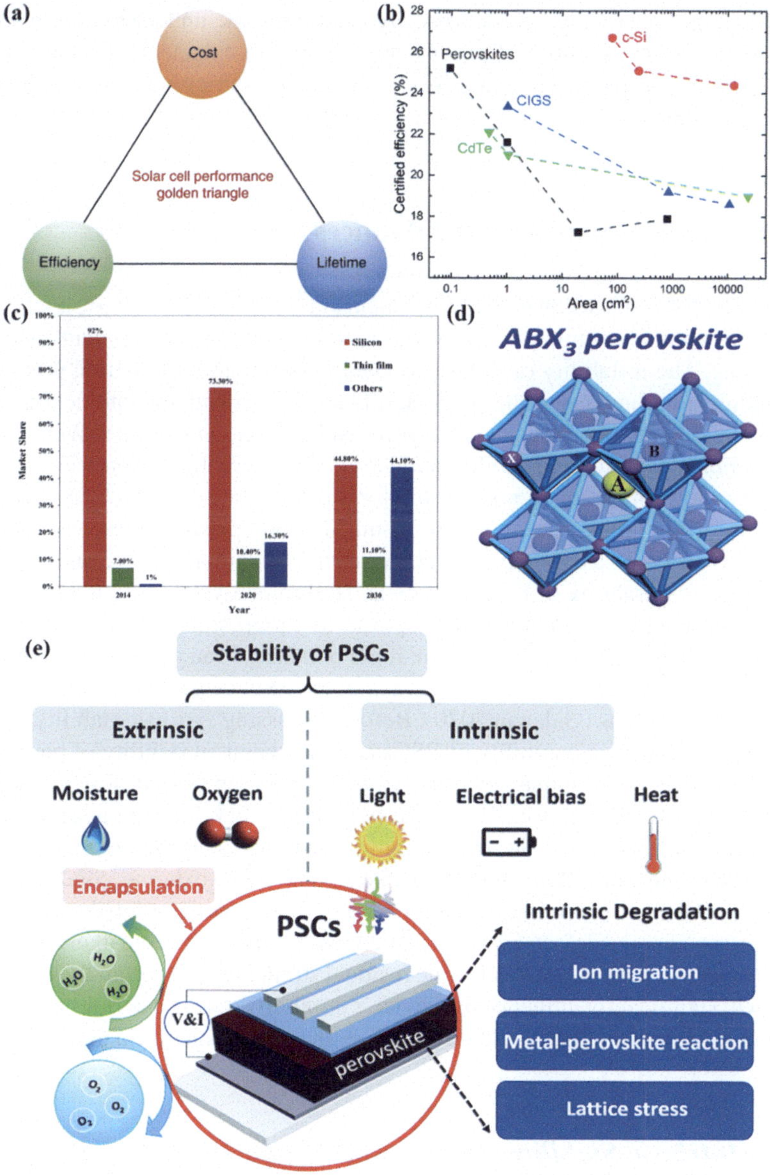

Fig. 2.1 **a** Schematic of Golden triangle for solar cell performance, Reproduced from Ref. Meng et al. (2018), Copyright (2018) with permission of Springer Nature. **b** Various certified PCEs of Si, CIGS, Perovskite, and CdTe Solar cells concerning device area, Reproduced from Ref. Werner et al. (2020), Copyright (2020) with permission of the Royal Society of Chemistry. **c** Global share of PV market (2014–30), Reproduced from Ref. Chowdhury et al. (2020), Copyright (2020) with permission of Elsevier. **d** Schematic of ABX_3 perovskite structure, and **e** Schematic demonstration of the extrinsic and intrinsic stability of PSCs under varying environmental and operational conditions, Reproduced from Ref. Cheng et al. (2021), Copyright (2021) with permission of the Royal Society of Chemistry

materials under operational conditions. By examining the underlying factors influencing the chemical stability of perovskites, we aim to enhance our understanding of the degradation pathways and develop strategies to improve the durability and longevity of PSCs.

2.2 Challenges and Recent Advances

Apart from all the successes and advancements in the PSCs technology, it still needs to catch up to global commercialization due to its instability under ambient conditions and toxicity. The instability of PSCs is related to the chemical reaction, perovskites are dealing with. Such reactions are generally associated with the interactions of the primary weak ionic bond, secondary hydrogen bonding, and van der Waals forces. In fact, ambient condition instability includes temperature, light, water, electric field sensitivity, etc., which are all responsible for the degradation of PSCs. An introductory study of the lattice structure and chemical bonds between the constituents of hybrid perovskites on a foundation level is crucial for a comprehensive understanding of the intrinsic durability of these materials. Hybrid perovskites are not chemically resilient and tend to break down in response to varied stimuli as the organic and inorganic entities interact through weak ionic bonds and van der Waals forces.

Figure 2.2 outlines the possible pathways for PSC degradation under ambient operational conditions (Galagan 2020). Before discussing extrinsic stability, we will emphasize the intrinsic durability of PSCs and the chemical stability of perovskites within a realistic environment. For instance, PSCs are employed in practical exterior applications with a powerful electric field, bright lighting, and a high working temperature of 55–65 °C. The fragile ionic bonds between the organic and inorganic ions in the perovskite may be broken under such extreme conditions, and because perovskites have a low activation energy for ionic migration, the fragmented charged ions (charged defects and dissociated ions) in them are more likely to migrate under an electric field. Additionally, the residual strain induced within the perovskite lattice will significantly influence the device's durability during exterior heat cycling, resulting in fatigue and steady deterioration of the device's performance.

2.2.1 Intrinsic Stability: Causes and Management

2.2.1.1 Ion Migration

Mizusaki et al. initially noticed the ion migration phenomena in PSCs, and they postulated that the movement of X-ion vacancies is the real cause of ionic conduction (Fig. 2.3a) (Yuan and Huang 2016). After Snaith et al. proposed electric field-induced ion migration in perovskites in 2014, ascribed as the main reason for *J*–*V* hysteresis in PSCs as depicted in Fig. 2.3b, this research issue gained considerable attention

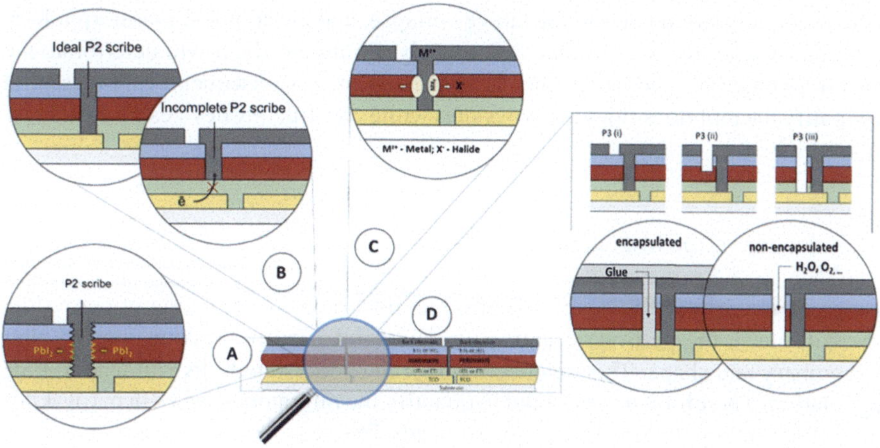

Fig. 2.2 Typical schematics of possible degradation mechanism in PSC modules. Reproduced from Ref. Galagan (2020), Copyright (2020) with permission of IOPscience

(Van Reenen et al. 2015). McGehee et al. explained that *J–V* measuring conditions (pre-bias, scan rate, and direction) significantly impacted the hysteresis in PSCs because of the photoinduced ion migration under an electric field. Huang et al. proposed the first concrete proof of ion migration in the device structured as ITO/ PEDOT: PSS/$MAPbI_3$/Au, which exhibits a fully reversed photocurrent orientation under reverse *I–V* scan attributed to the switchable ion migration concerning the + ve/−ve poling (Unger et al. 2014). These exciting results attracted researchers for further studies. Since then, ion migration has been effectively used to demonstrate anomalous effects such as photoinduced ionic conductivity and phase segregation, giant dielectric constant, etc.

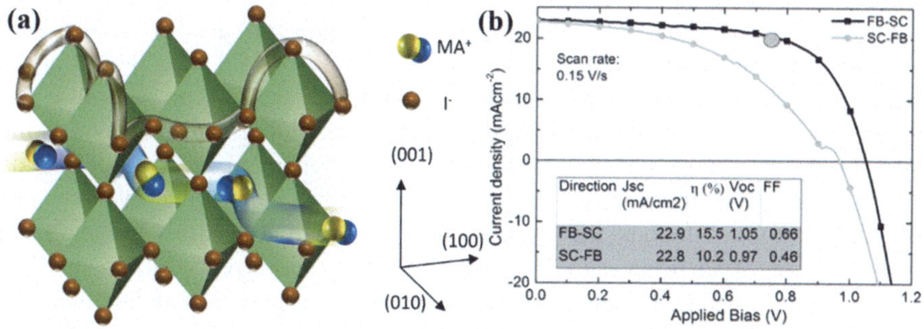

Fig. 2.3 **a** Schematic representation of ion migration in the general perovskite lattice, Reproduced from Ref. Yuan and Huang (2016), Copyright (2016) with permission of the American Chemical Society. **b** *JV*-hysteresis observation, one of the consequences of ion migration, Reproduced from Ref. Van Reenen et al. (2015), Copyright (2015) with permission of the American Chemical Society

Such ion migration leads to the lattice collapse, and the composition/morphology of the perovskite film will be changed entirely. Cumulatively, it will deteriorate the perovskite materials and reduce the PSC's performance. Consequently, it is essential to quantify the mobile entities in perovskites while constructing devices for extremely stable PSCs.

Quantification of Ion Migration

Whenever ion migration occurs in perovskite materials, the depletion of perovskite ions will cause the perovskite lattice to disintegrate and perovskite films to alter in chemistry and shape. This disintegration of the perovskite films will lead to PSC degradation. Therefore, it is essential to quantify the migratory species in perovskites while constructing devices for relatively steady PSCs.

He et al. presented a hole-only device, as depicted in Fig. 2.4a, in which the perovskite layer is laminated between ITO/PEDOT: PSS and MoO_3/Al electrodes. Owing to the comparable work functions of the ITO/PEDOT: PSS (5.0 eV) and MoO_3/Al (5.3 eV) electrodes, the hole-only device displays a flat band structure under equilibria circumstances in the initial state. The band flexion would vary if a voltage bias is introduced to the hole-only PSCs. For example, ITO/PEDOT: PSS and MoO_3/Al will act as the cathode/anode and vice-versa in the resultant diode when the $\pm$ 2 V bias was applied respectively (Fig. 2.4b). The charge density of migratory ions (D) collected at the contacts is obtained by integrating the relaxation current. The computed charge density peaked at $D = 284$ mC cm^{-2} at a voltage of 3.6 V, as depicted in Fig. 2.4c. Then, using the formula $N = D/qL$, one can calculate the mobile ion density (N) = 3.7×10^{19} cm^{-3} (Zhao et al. 2015).

Tsang et al. utilized a transient ionic current study (TICS) to investigate the migratory ion concentration and their impact on PSCs' performance. Figure 2.4d shows the device architecture was ITO/PEDOT: PSS/$MAPbI_3$/C_{60}/BCP, a *p–i–n* type architecture fabricated using two-step spin coating. At the same time, Fig. 2.4e depicts the TICS endorsed by ion-drift under V_{bi}, where bi stands for built-in. Initially, PSC was configured so that the mobile ions triggered by V_{bi} would be collected at the junctions. Secondly, an external voltage was applied to the PSC to neutralize V_{bi} and allow the accumulated ions to migrate to the other side. Finally, as soon as the external supply was removed, a transitory current was generated due to the collection of mobile ions at the interfaces. They concluded that this current mainly arises due to the motion of mobile ions; therefore, mobile ion density (generally ranging from 10^{16} to 10^{17} cm^{-3}) can be measured by integrating the ionic currents. Moreover, these mobile ions will act as a barrier in the charge carrier transportation and regulate the perovskite degradation by losing organic cations (Cheng et al. 2016).

Most recently, McGehee et al. replicated the device's electric field characteristic using a simulated band topology of PSCs. PSCs based on six distinct perovskite chemistries have been evaluated to assess the efficacy of their methodology, and each device has mobile ion concentrations that range from 7×10^{16} to 5×10^{17} cm^{-3} (Bertoluzzi et al. 2020) Other experimental and computational investigations have

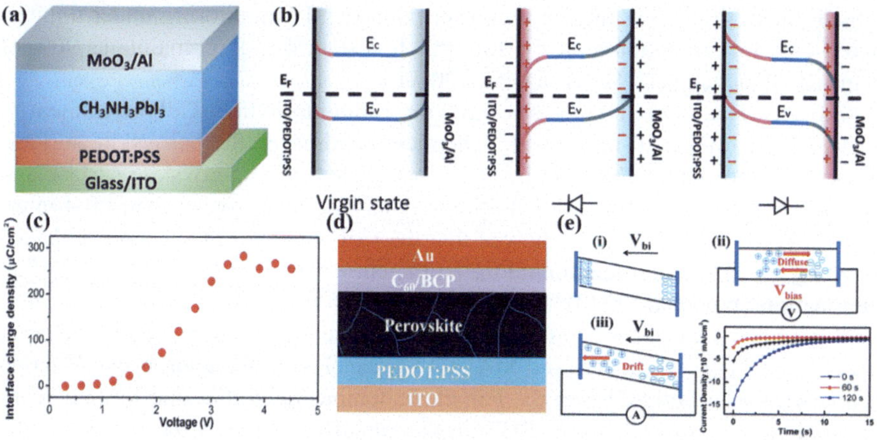

Fig. 2.4 **a** Schematic device architecture of polarity-switchable PSC, **b** energy band diagrams of the same polarity-switchable PSC under dark and illumination conditions, **c** estimation of the interfacial charge density at different bias-voltages, Reproduced from Ref. Zhao et al. (2015), Copyright (2015) with permission of the Royal Society of Chemistry. **d** PSC device architecture having different MAI loading, and **e** mobile ion-density measurement using transient ionic current investigation, Reproduced from Ref. Cheng et al. (2016), Copyright (2016) with permission of the Royal Society of Chemistry

shown mobile ion densities of roughly 10^{15}–10^{18} cm^{-3} in PSCs. Even though it has demonstrated that perovskites are highly defect tolerant and PCE reaches more than 25% even at a mobile ion concentration of 10^{17} cm^{-3}, uncontrolled mobile ion concentration inside the perovskite layer could bring additional undesirable ionic defects as well as hasten the deterioration of PSCs.

Scheme to Control Ion Migration

Based on the summary of the role of mobile ions in stabilizing PSC devices, it's clear that the intake of mobile ions causes PSCs to degrade permanently. Ion dissociation and ion migration in perovskite films in response to external stimuli are caused by the weak chemical bonds between the components. Therefore, (i) improving chemical bonding, (ii) neutralizing mobile ions, and (iii) decreasing mobile ion consumption are the three primary methods for reducing the adverse effects of mobile ions.

Solution-processed engineering improves the chemical bonding at the perovskite surface by creating a Pb-rich layer. The $FAMAPbI_3$ perovskite was originally spin-coated with a $Pb(SCN)_2$ solution by Han et al., who subsequently added a layer of chlorinated graphene oxide (Cl-GO) on top. The $Pb(SCN)_2$ and Cl-GO layers produce strong Pb–Cl and Pb–O connections to prevent the loss of the perovskite components beneath them. Following maximum power point (MPP) monitoring under uninterrupted 1-Sun illumination at 60 °C for 1000 h, a stabilized PSC maintains 90% of its initial efficiency of 21% (Wang et al. 2019b).

Similarly, Huang et al. suggested utilizing cadmium iodide (CdI_2) treatment to create a I^- rich layer to improve the chemical bonding on the perovskite surface. Figure 2.5a shows that CdI_2 can fill the I^- vacancies at the perovskite surface and immobilize the I^- ions because Cd–I ionic bonds are more robust than Pb–I interactions. Thus, the researchers showed that the CdI_2-treated PSCs exhibit a device efficiency of 21.9% and maintain a 92% efficiency over 1000 h of constant 1-Sun illumination (Wu et al. 2020). Another potential strategy to increase the intrinsic stability of perovskite semiconductors is to neutralize the fragmented ions and point defects (such as I^- vacancies and MA^+/FA^+ vacancies) in the appropriate location of the perovskite lattice. Recently, Pan et al. demonstrated that due to their identical ionic radius, super-halogen BH_4^- groups can effectively replace I^- ions in the perovskite lattice. In addition, as seen in Fig. 2.5b, the additional BH_4^- groups could create potent di-hydrogen bonds with MA^+ organic anions to maintain the perovskite lattice structure and increase the intrinsic stability of PSCs. Polymer additives were explored to improve the PSC's life-time as they set an obstacle for ion migration by gathering at grain boundaries of the perovskites (Xu et al. 2020b).

The chemical reactivity of I^- and Pb^{2+} against external stressors leads to start the disintegration of perovskite films, even though chemical engineering has been used to reduce ion migration in PSCs and thereby enhance the durability of PSCs as stated above. By including the Eu^{3+}–Eu^{2+} ion redox duo, Yan et al. used a creative technique to prevent the development of I^0 and Pb^0 in the perovskite layer. Accordingly, the Eu^{3+} ion introduced to the absorber layer can selectively oxidize Pb^0 to Pb^{2+} before being reduced to Eu^{2+}. The Eu^{2+} ion can simultaneously decrease I^0 to I^- and oxidize back to Eu^{3+}. In this way, the Eu^{3+}-Eu^{2+} ion pair functions as a "redox shuttle" to autonomously oxidize Pb^0 and reduce I^0, stopping the perovskite constituent from being consumed. Excellent device stability is shown by the related PSCs under standard illumination (Wang et al. 2019a). The aforementioned research clearly shows that one of the most effective ways to increase the practical stability of PSCs is to reduce ion migration and prevent ion consumption of perovskite films while the device is operating.

2.2.1.2 Metal-Perovskite Interactions

Under solar irradiance, perovskite layers initially absorb photons and produce electron–hole pairs; these are then differentiated and captured by the metal electrodes. An ohmic contact between the perovskite layer and the metal electrode was chosen to lower the contact resistances, enabling charge transport at the interfaces to achieve effective charge carrier collection. As a result, metal electrodes with lower work functions, like Al and Ag, are frequently used in *p–i–n* shaped PSCs, but higher work function metal electrodes, like Au, are preferred in *n–i–p* structures (Saris et al. 2019). Despite the excellent conductance of the metal electrodes, it has been found

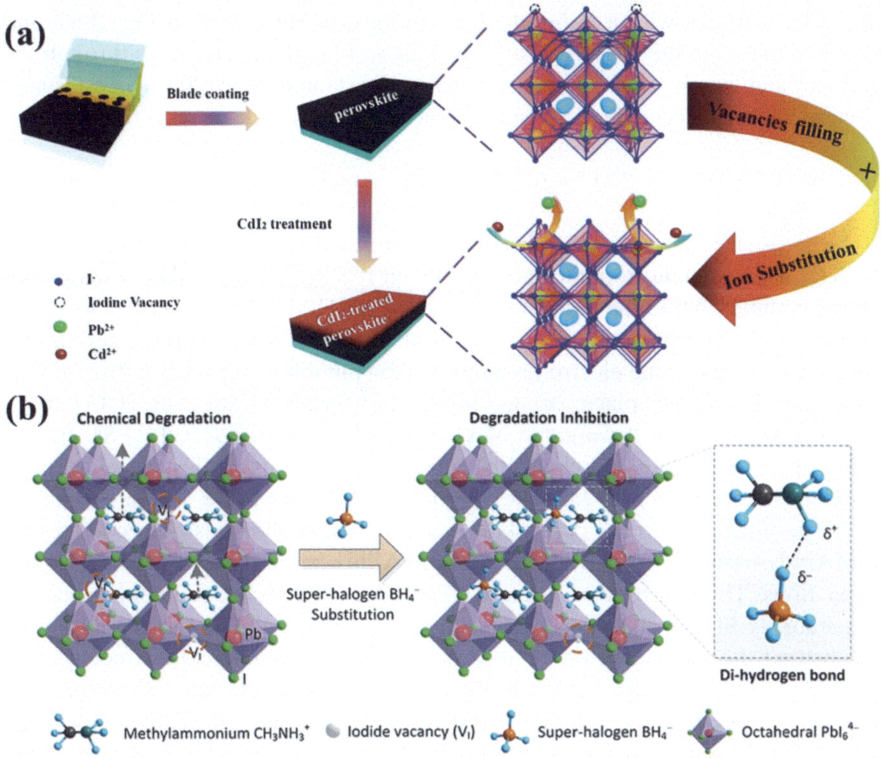

Fig. 2.5 **a** Typical blade-coated and CdI_2-treated perovskite film, Reproduced from Ref. Wu et al. (2020), Copyright (2020) with permission of the Americal Chemical Society. **b** Schematic representation of di-hydrogen bond formation in between super-halogen and MA^+ ions, Reproduced from Ref. Xu et al. (2020b), Copyright (2020) with permission of the Americal Chemical Society

that the rapid migration of metallic particles into the perovskite film and their reactivity with perovskite constituents significantly influenced the protracted stability of the device. Because this metal-perovskite reaction-persuade perovskite breakdown cannot be remedied by encapsulating technology, it is regarded as the intrinsic stability problem of PSCs.

Metal-Perovskite Interaction Mechanism

It is commonly recognized that among the deterioration mechanisms in organic solar cells is the corrosion of metal electrodes. Snaith et al. noted that whenever the Ag electrode interacts with perovskite materials, AgI will form and cause corrosion of the electrode. However, there needs to be concrete proof that their work involved an Ag-perovskite reaction. As stated above, stopping the ion migration completely in the perovskite films is challenging (Leijtens et al. 2013). Ma et al. created PSCs using Al electrodes with a *p–i–n* device topology, as illustrated in Fig. 2.6a, and illustrated that both the $CH_3NH_3^+$ and I^- ions would effectively move through the interfaces and transit via its charge transfer layers (Guo et al. 2020). The estimated redox interactions occurring between Al-perovskite interfaces were as follows:

$$6CH_3NH_3PbI_3 + 2AI \rightarrow 6PbI_2 + 1AlI_3 + 6CH_3NH_3 \uparrow +3H_2 \uparrow \tag{2.1}$$

The step-wise Ag-perovskite interactions are evidently shown in Fig. 2.6b, in which (i) water content from the surrounding air reaches the perovskite thin coating through pinholes in the hole transfer layer (spiro-MeOTAD); (ii) the perovskite disintegrated into MAI, PbI_2, and HI; (iii) the MAI and HI iodine-containing derivatives relocate toward the metal electrodes again via the pinholes; (iv) surface diffusion of iodine-complexes takes place; (v) and finally AgI formed (Kato et al. 2015). Yang et al. investigated how the organic MA^+ and FA^+ cations impact the PSC device deterioration to better comprehend the metal-perovskite dynamics. In addition to the I^- ions, as seen in Fig. 2.6c, the Ag ions from the electrode were also susceptible to permeating into the perovskite material. The authors also discovered that the $MAPbI_3$-based PSCs are less stable than $FAPbI_3$-based PSCs and offered many explanations. The first reason is that MA^+ and FA^+ cations are different in size. A comparatively bigger organic cation, such as FA^+, will stabilize the crystal lattice and obstruct the movement of I^- ions because these organic cations occupy the center $[PbI_6]^{4-}$ site. Second is the low dipole moment of FA^+ ions. Therefore, the energy barrier is comparatively very high to rotate them. Lastly, the activation energy required for I^- ion migration is very low in $MAPbI_3$ compared to $FAPbI_3$ (Zhang et al. 2017).

One of the potential replacements for Al or Ag as a metal contact for PSCs is the noble Au. Au is not a reliable contact material for PSCs, despite numerous investigations showing how PSCs employing Au as an electrode have longer longevity than devices using Ag electrodes. Au atoms and ions have been shown to diffuse into perovskite films via organic hole transfer layers and interact with the material there. High thermal stress and sustained light irradiation can also facilitate these Au-perovskite reactions (Liang et al. 2018). For instance, Tarasov et al. thoroughly investigated the nature of Au electrode deterioration resulting from its interaction with iodine-based perovskites (Shlenskaya et al. 2018). However, Rand et al. discovered the generation of Pb^0 species in a different investigation at the $MAPbI_3$/Au contacts, not at the PbI_2/Au contact, indicating that the organic MA^+ is involved in the Au-perovskite interactions. According to the literature mentioned above, MA^+ and I^- species may act as catalysts for the deterioration of Au contacts (Kerner et al. 2019). As a result, additional, more durable, reasonably priced conductive materials will be needed as electrodes for the widespread use of perovskite solar technology.

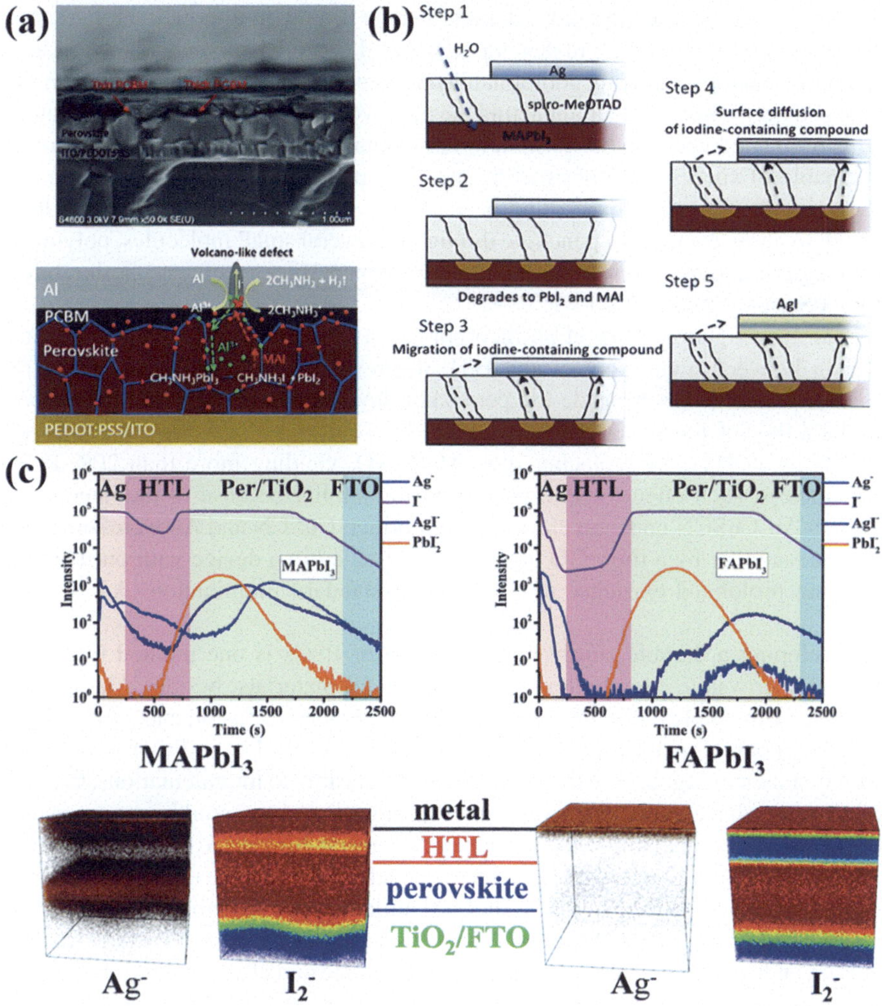

Fig. 2.6 **a** SEM image of *p–i–n* structured PSC with Ag contact, Reproduced from Ref. Guo et al. (2020), Copyright (2020) with permission of WILEY-VCH Verlag GmbH & Co. KGaA, Weinheim. **b** Possible degradation pathways arising from metal (Ag)-perovskite interaction, Reproduced from Ref. Kato et al. (2015), Copyright (2015) with permission of WILEY-VCH Verlag GmbH & Co. KGaA, Weinheim. **c** Elemental distribution using TOF–SIMS profiling, Reproduced from Ref. Zhang et al. (2017), Copyright (2017) with permission of the Royal Society of Chemistry

Scheme to Control Metal-Contact-Induced Degradation

Several techniques have been used in the line of work to reduce the deterioration of PSCs brought on by the infamous metal-perovskite interactions, including (i) introducing intermediate compact layers blocking between the metal electrode and perovskite film, (ii) locating more durable metals to find an alternative for Al, Ag, and Au, and (iii) continuing to develop non-metal conductive contacts for PSCs. It is simple to advocate placing additional layers to prevent direct contact between the metal and perovskite without impacting the photovoltaic efficiency. First, the coating that is implanted needs to be resistant to both metal and perovskite. Second, it must be capable of successfully obstructing metal atom/ion diffusion or perovskite ion migration. Thirdly, the thin coating needs to be a charge transfer layer or skinny enough to allow excitons to penetrate through it. Several small molecules, polymers, and inorganic compounds were suggested as a compact layer to enhance the durability of the PSCs.

In an investigation, Park et al. incorporated a poly(3-hexylthiophene)-polymer layer in between metal contact and HTL to prevent metal-perovskite interaction and moisture ingression inside the perovskite layer, which significantly enhanced the durability of the device (Kim et al. 2017). Gratzel and co-workers employed CuSCN as an HTL by replacing spiro-MeOTAD, yielding more than 20% PCE. Nevertheless, such effective PSCs showed low stability because of the interaction between Au-CuSCN under an electric field. Further, CuSCN and Au were separated from one another by a thin rGO layer, which resulted in a device with outstanding operating prolonged existence at 60 °C under standard illumination (Arora et al. 2017).

Developing a durable sturdy metal contact for PSCs is one method to reduce electrode-perovskite degradation. Numerous metal materials, besides Al, Ag, and Au, have been employed as PSC contacts. The computed diffusion energy threshold of several contact materials (Cu, Ag, Au, Cr, Mo, W, Ni, Pd, and Co) in $MAPbI_3$ PSCs is shown in Fig. 2.7a (Ming et al. 2018). According to the calculations, Cr, Mo, and W metals have higher diffusion barriers, suggesting they could be used as metal contacts in PSCs. In addition to the stability, the resistivity and work function of the metal are additional significant aspects to consider for an ideal device electrode in the physical realm. According to Fig. 2.7b, Cu is anticipated to be stable for 170 years at 27 °C and 22 years at 40 °C based on information gathered at laboratory time frames. Cu is a viable contact material for building very stable and effective PSCs because it can create PSCs with over 20% efficiency and excellent stability (Zhao et al. 2016b).

Creating non-metal contact materials is an effective way to prevent device degradation brought on by metal-perovskite interactions. Various non-metal contact materials have been created for PSCs to support device stability, notably carbon- and TCO-based contacts. A PSC without an HTL that uses carbon as a non-metal electrode has been designed by Han et al. A porous carbon sheet was initially applied as a counter electrode on a double ETL of mesoporous TiO_2 and ZrO_2 that served as a scaffold for the perovskite. The pores in the carbon electrode, TiO_2, and ZrO_2 scaffold were filled in by pouring the perovskite precursor from the top. In the ambient environment and standard illumination, the cell had a verified PCE of 12.8% with stability of 1000 h (Mei et al. 2014). McGehee et al. employed aluminum-doped zinc oxide nanoparticles as an interlayer to safeguard the perovskite film when depositing the ITO on top. The outcomes showed that the ITO electrode-encapsulated PSCs prevented metal-perovskite-induced device deterioration, offering a potential

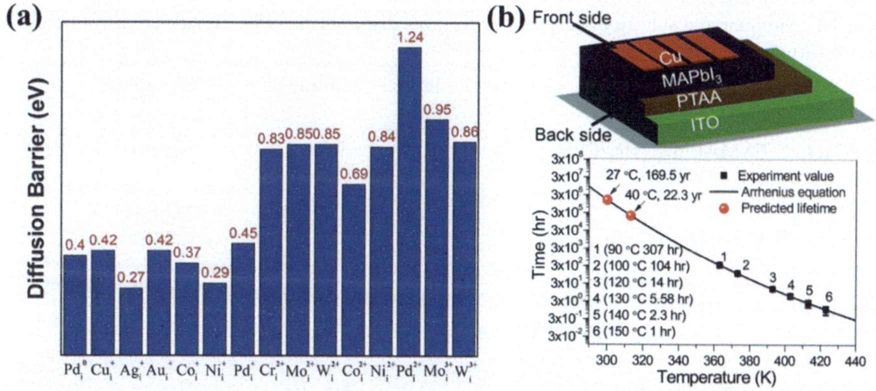

Fig. 2.7 **a** Estimated diffusion barrier of various metals interstitial defects in $MAPbI_3$, Reproduced from Ref. Ming et al. (2018), Copyright (2018) with permission of WILEY-VCH Verlag GmbH & Co. KGaA, Weinheim. **b** PSC device with Cu electrode and their lifetime, Reproduced from Ref. Zhao et al. (2016b), Copyright (2016) with permission of the Royal Society of Chemistry

method to produce stable PSCs with more than 25 years of stability (Bush et al. 2016). Considering diverse endurance evaluation regimes, Table 2.1 highlights the durability of PSCs from various perovskite chemistries and device architectures. Consequently, to obtain very stable PSCs for upcoming practical applications and operational circumstances, we advise the development of highly efficient ITO or other strong TCO electrodes. We still urge perovskite researchers to continue making many more initiatives on this subject even though PSCs with translucent TCOs have fallen underneath their equivalents with opaque contact electrodes in PSCs performance owing to optical absorption loss occurring due to the translucent TCO top electrode.

2.2.1.3 Residual Strain

The physicochemical characteristics and long-term durability of several semiconducting materials, including germanium and GaAs, have been discovered to be impacted by lattice strain. Perovskites, being of a polycrystalline character, are similarly subject to lattice strain effects. Given that the perovskites generally have a thermal expansion coefficient of 1.57×10^{-4} K^{-1}, which is six times greater as compared to soda lime glass, it may be possible that residual strain comes into the picture in the course of film formation upon thermal annealing (Jacobsson et al. 2015). In particular, the foundation would prevent the perovskite from contracting and creating lattice strain during cooling immediately after thermal annealing. Additionally, the PSCs experience cycling compressive stress and tensile stress due to temperature variations when operating in realistic day/night cycles. Hence, such lattice tension inside the perovskite coating would result in progressive fissures in PSCs and chemical deterioration.

Table 2.1 Summarized stability analysis with various distinguished PSC device architectures and their performance

PSC structure	PCE (%)	Stability	References
FTO/TiO_2/CsFAMAPb$I_{3-x}Br_x$/CuSCN/rGO/Au	20.4	> 95%, 1000 h	Arora et al. (2017)
FTO/TiO_2/$Cs_5(MA_{0.17}FA_{0.83})_{95}Pb(I_{0.83}Br_{0.17})_3$/spiro-MeOTAD/Au	21.1	85%, 250 h	Saliba et al. (2016b)
FTO/TiO_2/$Rb_5(Cs_5(MA_{0.17}FA_{0.83})_{95})_{95}Pb(I_{0.83}Br_{0.17})_3$/spiro-MeOTAD/Au	21.6	95%, 500 h	Saliba et al. (2016a)
ITO/C_{60}-SAM/SnO_x/$PC_{60}BM$/$FA_{0.83}$/$MA_{0.17}Pb_{1.1}Br_{0.5}I_{2.8}$/PDCBT/Ta-$WO_x$/Au	21.2	95%, 1000 h	Hou et al. (2017)
ITP/PTAA/$Cs_{0.05}FA_{0.81}MA_{0.14}PbI_{2.55}Br_{0.45}$/$C_{60}$/BCP/Cu	21.1	96.8%, 1200 h	Yang et al. (2019)
ITO/NiO_x/$MAPbI_3$/PCBM:C_{60}/ZrAcac/Cu	20.1	70%, 500 h	Zhang et al. (2018)
FTO/NiO/$(FA_{0.83}MA_{0.17})_{0.95}Cs_{0.05}Pb(I_{0.9}Br_{0.1})_3$/PCBM/BCP/Cr/$Cr_2O_3$/Au	19.8	95%, 1885 h	Bai et al. (2019)
FTO/SnO_2/PMMA/$Rb_5Cs_{10}FAPbI_3$/PMMA/spiro-MeOTAD/Au	18.7	93%, 1000 h	Turren-Cruz et al. (2018)

Strain-Induced Degradation Mechanism

The discrepancy in the thermal expansion coefficients of perovskite and ITO/FTO causes considerable residual tensile strain in the perovskite film formation process via spin coating, as displayed in Fig. 2.8a (Rolston et al. 2018). At the interface of perovskite/ITO/FTO under high enough temperatures, glass substrates were not good enough to allow proper lattice expansion of the perovskite film. The XRD approach characterizes the lattice strain by measuring the peak shift as residual strain directly correlated to the unit cell properties (Zhao et al. 2017). Zhu et al. used depth-dependent grazing incident X-ray diffraction (GIXRD) characterization to clarify the gradient growth of residual strain in the vertical orientation of mixed-halide perovskite films. It is assumed that gradient thermal stress resulting from film manufacturing and/or possible composition inhomogeneity are the causes of the residual strain distribution that was observed. In order to adjust the type of strain (compressive or tensile) and gradient of strain in the perovskite film, a simple method was developed. Using this technique, it was possible to determine that the residual strain gradient had a major effect on the hole carrier dynamics throughout the perovskite solar cell, as depicted in Fig. 2.8b–d. First-principle simulations revealed that the strain gradient causes the perovskite absorber's valence bands to flex, which in turn affects the dynamics of the interfacial holes. A notable improvement in hole carrier transport and extraction across the perovskite absorber/HTM interface was obtained via strain engineering, which reduced the strain gradient of the perovskite layer. As so, the perovskite solar cell that was tuned achieved a validated PCE of 20.7%. This work advances our understanding of the composition, structure, and property connection within the halide perovskite system in addition to demonstrating the impact of residual strain on the optoelectronic properties of the halide perovskite film (Zhu et al. 2019).

To get a strain-free XRD plot, Huang et al. investigated a single-crystal $MAPbI_3$ and found that thermal annealing was the main culprit behind the formation of residual strain. There is a clear peak shift in annealed single-crystal $MAPbI_3$, while non-annealed samples have shown identical peaks. While in-plane XRD offers details of the spacing between parallel planes, out-of-plane XRD mainly reflects perpendicular plane spacing. As a result, horizontal tensile strain is indicated by the in-plane peak shifting to a relatively low angle. In contrast, lower plane separation induced by vertical compressive strain is indicated by the out-of-plane XRD peak changing to a greater angle (Zhao et al. 2017). To examine the lattice strain, PL and time-resolved-PL investigations were also used in addition to XRD analysis.

Scheme to Control Strain Prompted Degradation

According to the description above, thermal treatment is the primary factor in the origin of the tensile strain. The lattice widening under tensile strain significantly affects the crystal lattice variables, which downturns the bond strength. Further, this will lead to defect creation, thermal ion decomposition, and migration. The algorithm can be used to forecast the remaining lattice strain due to temperature variation as follows:

$$\sigma_{\Delta T} = \frac{E_p}{1 - \nu_p}(\alpha_s - \alpha_p)\Delta T \tag{2.2}$$

where E_p, ν_p, α_s, α_p, and ΔT are Young's modulus, Poisson's ratio, thermal expansion coefficient for substrate and perovskite, and temperature difference, respectively. Accordingly, low-temperature thermal treatment is the simplest way to lessen residual strain during the production of perovskite and crystallization (Wang et al. 2018). However, a bath switchover approach is being described to create perovskite films at relatively lower temperatures; high-temperature treatment of perovskite films is always necessary to produce an excellent perovskite film for highly efficient PSCs.

To diminish the residual strain induced on the perovskite films, Sargent et al. inserted a hole transfer film with a very large thermal expansion coefficient between the perovskite film and glass only to minimize the expansion discrepancy (Liu et al. 2019). In addition, Fig. 2.9a, b display the usual thermal treatment temperatures to produce high-quality and defectless perovskite films.

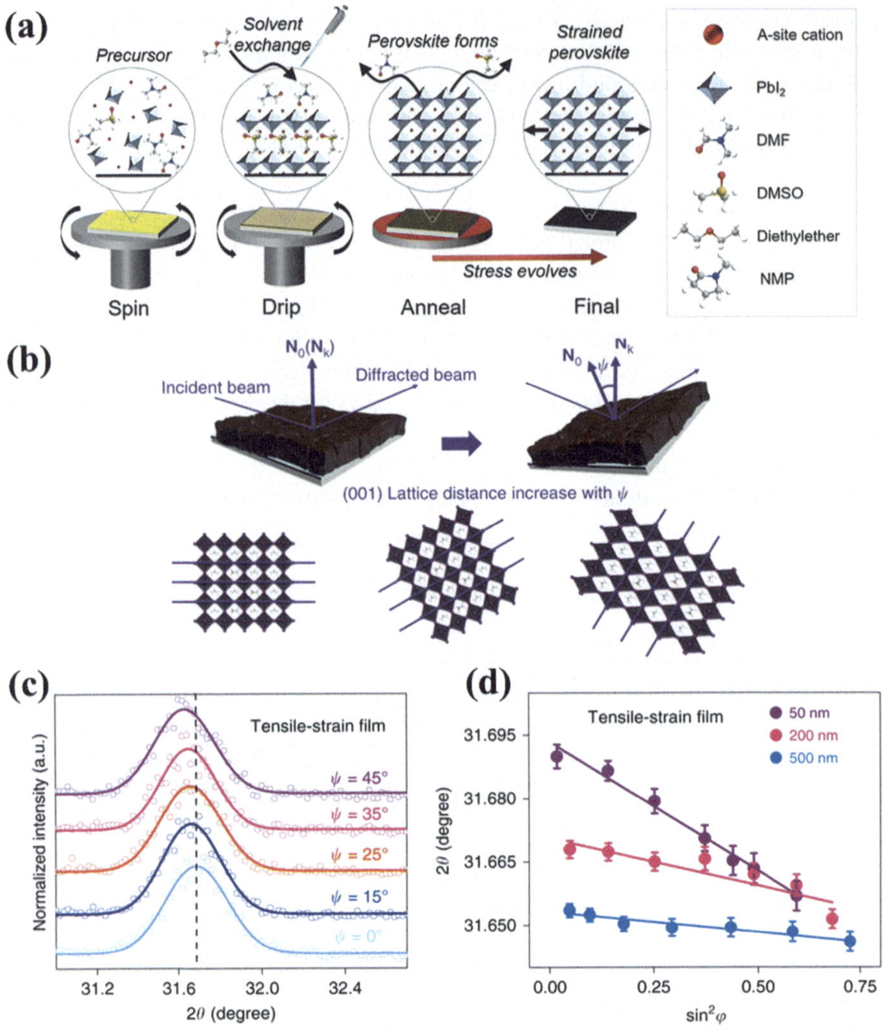

Fig. 2.8 **a** Typical schematics of the residual stress formation during PSC fabrication, Reproduced from Ref. Rolston et al. (2018), Copyright (2018) with permission of WILEY-VCH Verlag GmbH & Co. KGaA, Weinheim. **b** Schematic representation of residual strain distribution and determination of lattice strain by considering N_0 as normal to the sample, N_k as the diffraction vector, and ψ as a tilt angle. **c**, **d** GIXRD pattern for the strained film at different tilt angles @50 nm depth and residual strain distribution attributed at the different depths (50, 200, and 500 nm) in tensile strained film, Reproduced from Ref. Zhu et al. (2019), Copyright (2019) with permission of Springer Nature

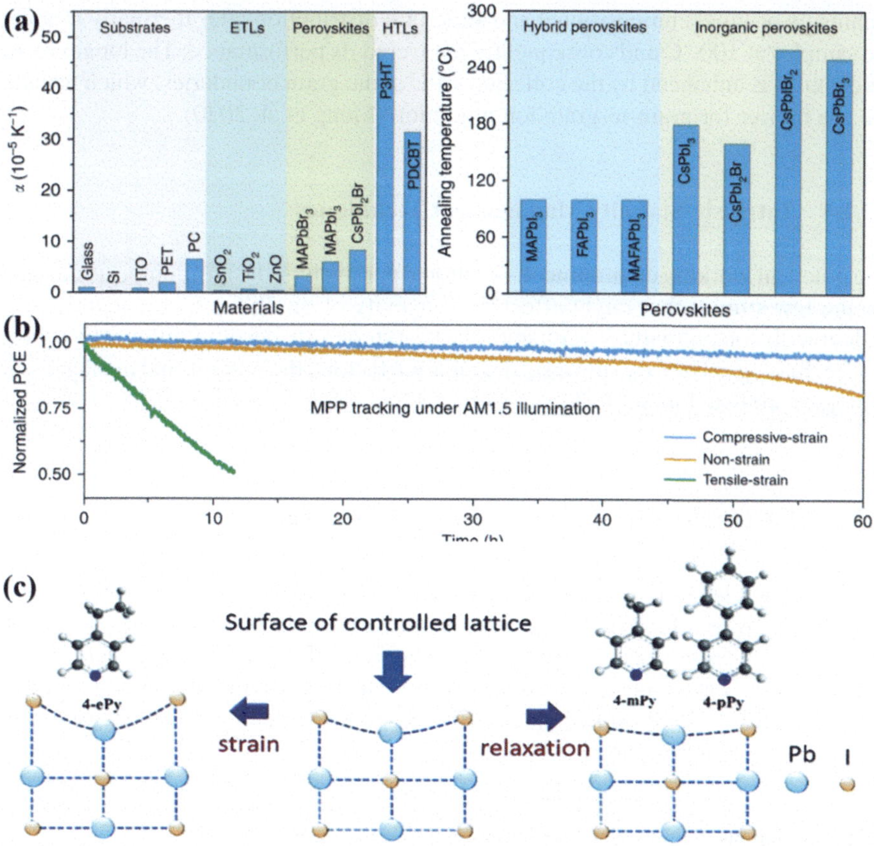

Fig. 2.9 **a** Thermal expansion coefficient of different PSC layers and general annealing temperature of different perovskite films, and **b** MPP tracking under AM1.5 illumination, Reproduced from Ref. Cheng et al. (2021), Copyright (2021) with permission of the Royal Society of Chemistry. **c** Strain relaxation process via pyridine derivative, Reproduced from Ref. Xu et al. (2020a), Copyright (2020) with permission of Elsevier

Although such significant temperature variations cannot be avoided entirely during perovskite creation, a strain-management strategy can be used by sandwiching the perovskite layer with low or high-α materials that require the induced strain. Pyridine derivatives can be used as reinforcements within perovskite films to increase the PSCs' efficiency and longevity. As depicted in Fig. 2.9c, a cross-linked ligand of pyridine derivatives will minimize the defect states and improve the lattice stress relaxation, resulting in exceptional 21.12% PCE and sustaining long-term reliability for over 1000 h (Xu et al. 2020a). The stress relaxation mechanism can be further investigated in flexible PSCs by employing self-healing polyurethane (s-PU) on some stretchable substrates. It is revealed that initially, s-PU entirely covered the grain boundary regions, but it will destroy the whole dynamics during bending. Further, it regains its original compositional and structural distribution after thermally treating the samples at 100 °C and consequently recovered its performance. The longevity of perovskites is enhanced by the collected s-PU at the grain boundaries, which usually act as a barrier for grain-to-grain ion migration (Meng et al. 2020).

2.2.1.4 Intrinsic Stability Management Techniques

The practical working environment, functional parameters, design, composition, etc., are the few stimuli that might affect the longevity of the PSCs. Consequently, it is undoubtedly too premature to conclude the best strategy to develop sufficiently robust PSC technology. The main intrinsic degradation factors, their cause, and management strategies are listed in Table 2.2.

2.2.2 Extrinsic Stability: Causes and Management

The commercialization of perovskite-based PV technology is considerably constrained by several extrinsic problems, notwithstanding the technology's quick advancement in terms of device efficiency. The short lifespan, adverse effects of lead leakage on the environment, and a lack of field-testing examination are only a few of the difficulties in introducing perovskite-based PV technology to the market that will be covered in this part.

Table 2.2 Various intrinsic in-stability causes and their management strategies

Factor	Degradation pathways	Stability management
Electrical bias	✓ prompted ion dis-integration and migration ✓ field-prompted phase separation	✓ Mixed 2D/3D perovskite structures ✓ Use of polymer additives
Residual stresses	✓ Fatigue in the perovskite film ✓ Induced ion migration as chemical bonds get enfeebled	✓ In situ stress management ✓ Additive/interfacial engineering
Metal electrode	✓ Metal–perovskite interactions	✓ Preferably use non-metal electrode ✓ Interfacial engineering to manage the metal-perovskite direct contact
Perovskite composition	✓ Erratic nature of MA^+, FA^+, etc. ✓ Prompted defect states	✓ Encapsulation ✓ Defect controlling
Interlayer materials	✓ Volatility of organic charge transfer layers	✓ Robust inorganic charge transport layers

2.2.2.1 Lead (Pb) Leakage

Before perovskite-based PVs can be commercialized, the toxic nature of Pb that has a negative impact on humans and the natural world must be resolved. For instance, if the perovskite PV modules are harmed by rain, the Pb stored inside the cells may leak out and contaminate the soil and groundwater. According to a recent study by Dedecker et al., the quantity of Pb used in PSC modules to supply the USA and the world with electricity is 160 and 1000 tons per year, respectively. Comparatively, lead acid batteries and electronic solders each require 6200 and 1,000,000 tons of Pb annually (Dedecker and Grancini 2020). Here, the perovskite layer thickness was considered to be between 300 and 500 nm to estimate the Pb content. From the results, it was predicted that there are 0.04–0.07 mg of Pb in a 1 cm^2 perovskite cell, or 0.4–0.7 gm of Pb in 1 m^2 of perovskite-based PV modules (as the density of $MAPbI_3 = 4.16$ gm cm^{-3}) (Zhou et al. 2015). Although the amount of Pb in PSC modules is relatively low compared to other Pb-based technologies (such as lead-acid batteries), stopping Pb leakage from PSC modules is still a top priority in this industry.

Figure 2.10 depicts two potential Pb leakage routes from PSC devices: perovskite in water and perovskite on fire. Figure 2.10a, b illustrate how Han et al. placed the PSC devices in a fire simulation setup to assess Pb permeability when the ensembles were exposed to a fire outbreak (Bi et al. 2019). The temperature within the fire simulation setup was set at 760 °C with a continuous flow of clean air to simulate a structural fire. As seen in Figure 2.10c, d, the PSC modules' cover glass deteriorated, and the devices disintegrated when exposed to such a high temperature. After the perovskite PV devices accidentally come into contact with fire, XPS examination shows that the ultimate products are PbO and PbO_2, respectively. According to the theory, the perovskite first breaks down into PbI_2 at high temperatures, and subsequently, PbI_2 is oxidized to PbO and PbO_2, as seen in other reports (Jiang et al. 2019). In addition, the breakdown of perovskites with contact with water is depicted in Fig. 2.10e–g, where the $MAPbI_3$ perovskite initially transforms into a colorless monohydrate phase ($MAPbI_3 \cdot H_2O$) before continuing to decompose into a yellowish PbI_2. To examine changes brought on by water in $MAPbI_3$ thin films, Yang et al. used grazing-incidence wide-angle x-ray scattering (GIWAXS). Real-time data acquisition was made possible by the rapid 2D area detector and high-brilliance synchrotron source. Before the decomposition to PbI_2 and under high RH conditions, the measurement revealed a consolidation stage in the breakdown pathway (Yang et al. 2015).

Further, pXRD experiments lend support to this hypothesis. PbI_2 and aqueous MAI were shown to be the final breakdown products by Zhao et al. (using solid–liquid extraction), supporting the PbI_2 formation first noted by Yang et al. The ^{207}Pb solid-state nuclear magnetic resonance (NMR) spectroscopy successfully detects minute alterations in the local chemical and coordination context and provides additional support for the monohydrate phase (Zhao et al. 2016a). Askar et al. discovered chemical changes in the ^{207}Pb NMR spectra of $MAPbI_3$ after exposing it to 80% RH, which point to the monohydrate phase if heated, causing the sample to break down to PbI_2 (Askar et al. 2017).

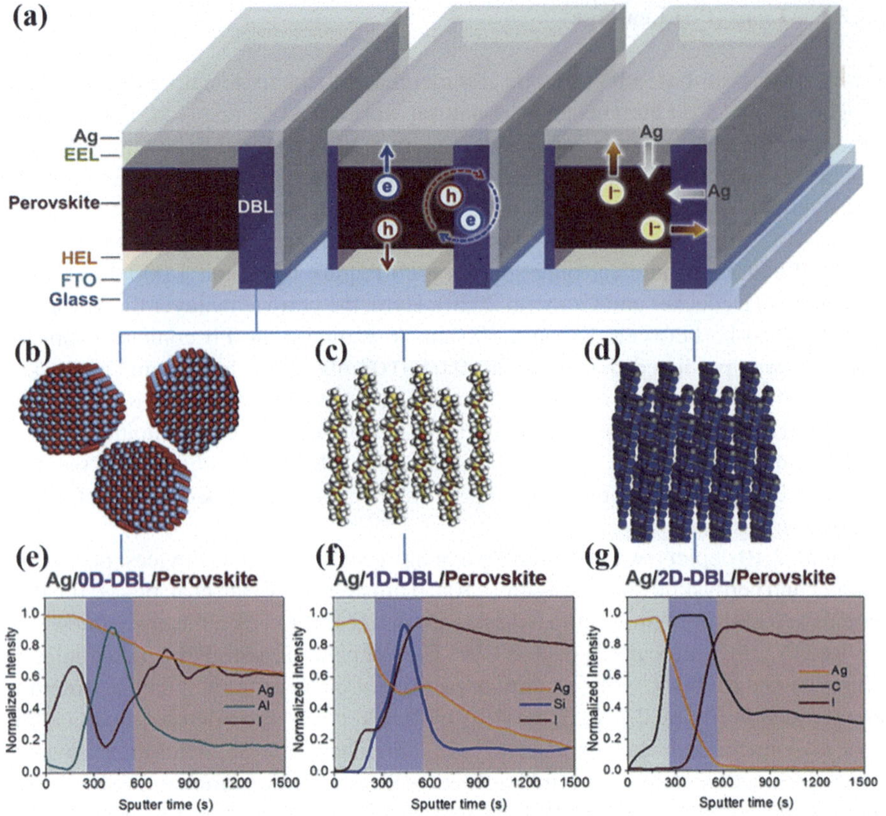

Fig. 2.10 **a** Schematic illustration of the vertical and parallel diffusion in the PSC module, **b** Insertion of 0-D Al_2O_3, **c** 1-D PDMS, and **d** 2-D *g*-C_3N_4 diffusion barrier layers, **e**–**g** The depth profiling of Ag and I with Al, Si, and C signals, Reproduced from Ref. Bi et al. (2019), Copyright (2019) with permission of Elsevier

Numerous techniques have been used to examine how hydration affects the optoelectronic characteristics of the perovskite, including steady-state and time-resolved in situ photoluminescence (PL) spectroscopy and in situ ultraviolet photoelectron spectroscopy (UPS). $MAPbI_3$ films were subjected to a humid nitrogen flow (80 $\pm$ 5% RH) when taking PL emission observations. The PL intensity was increased when the $MAPbI_3$ films were subjected to moisture because water molecules started to passivate the top surface by forming hydrogen bonds with uncoordinated iodine atoms (Song et al. 2018a). As shown by in Situ UPS measurements, higher RH causes a thin layer of water to form on the surface, which raises the free electron density there and causes n-type doping.

Pb leakage from perovskite-based PV systems can be reduced using two primary techniques. The first involves utilizing an encapsulating strategy to keep perovskite devices out of the extrinsic factors. In contrast, the second involves employing a material chemistry approach to capture Pb ions once the perovskites begin to break down. Jiang et al., for instance, conducted a thorough investigation on the Pb leakage rate from mechanically damaged perovskite PV modules using various encapsulation methods in different weather conditions. They showed that epoxy resin encapsulation, superior to ultraviolet-cured resin encapsulation, can effectively prevent Pb leakage (Jiang et al. 2019). Li et al. used a chemical technique to capture Pb when the PSC devices were broken and exposed to water. Two Pb-absorbing layers surround the entire perovskite device. A DMDP layer with good Pb absorption was coated on the glass substrate side, and on the top electrode side, the whole perovskite device was covered with an EDTMP-PEO blended layer (Li et al. 2020). Alternative Pb-trapping materials, like a 2D-conjugated metal–organic framework with thiol-functionalization, have also been used as ETL in PSCs to increase device efficiency while lowering Pb leakage. Pb-absorbing resins were also added to perovskite layers to collect the Pb ions when the devices were destroyed. Pb-absorbing resins have been shown to minimize Pb leakage and passivate flaws in defective PSC modules to enhance device performance. The above-mentioned studies offer realistic solutions to Pb leakage problems that arise when harmed PSC modules are submerged in water.

Unfortunately, a fire mishap could damage these encapsulations and Pb-absorbing layers, making it impossible to stop Pb leaking from the perovskite modules. New innovative and successful solutions are required to address the Pb leakage difficulties in the PSC industry. A different strategy for developing environmentally friendly perovskite-based PV technology is to collect and reuse the Pb leftovers left over from making the perovskite during manufacturing.

2.2.2.2 Humidity-Induced Degradation

The environmental component that most significantly affects PSC performance is moisture. It has been demonstrated that a tiny amount of moisture (either atmospheric humidity or water added to the perovskite ink) during the synthesis of the perovskite film improves film quality, resulting in larger grain sizes and lesser grain boundaries. However, the perovskite lattice is severely harmed by extended exposure to water vapor and high relative humidity (RH). The perovskite decomposes over days when unencapsulated $MAPbI_3$ films are left exposed to ambient air (RH = 40%, T = 22 °C), as seen by a visible coloration from black to yellow (Wang et al. 2017).

The phase changes that take place during degradation were well understood with the X-ray scattering experiments. Nevertheless, pXRD offers no insight into the sample's amorphous elements, and characteristics involving weakly scattering (low Z) components are frequently hidden. To get around these problems and learn more about water uptake in the $MAPbI_3$ films, Schlipf et al. employed in situ grazing-incidence small-angle neutron scattering (GISANS) studies using heavy water (D_2O). Their studies reveal that the outermost layer of perovskite crystallites primarily absorbs water; if the RH is mild ($\leq$ 58%), the whole moisture limit can be taken in. The $MAPbI_3$ exhibits structural changes and produces the monohydrate phase at higher RH ($\geq$ 73%). They noticed signs of the di-hydrate stage at higher RH ($\geq$

93% RH) (Schlipf et al. 2018). The Yoneda peak position indicates that the film may contain equal to 50% (v/v) water at this point. Ex situ pXRD failed to detect the hydrate phases, demonstrating their ephemeral nature at lesser water partial pressures.

According to the GISANS results, $MAPbI_3$ can still absorb water at low RH levels. As a result, morphological rather than structural changes are probably the initial indications of breakdown. After just a few hours of exposure to an RH of 80%, $MAPbI_3$ films exhibit a rise in surface roughness and a reduction in grain size, as shown in Fig. 2.11, according to in situ scanning force microscopy (SFM) (Li et al. 2016) As no additional morphological changes were seen after drying, it was determined that this was an irreversible process. Leguy et al. found similar roughness increases and grain size reduction in the ex situ atomic force microscopy (AFM) investigation. Ex situ scanning electron microscopy (SEM) data contends that water molecules enter the perovskite layer along the spaces between grains, providing additional evidence. As a result, decomposition begins at the grain borders and moves into the grain core. This appears as a tapering of the grain at high humidity levels, and after extended exposure, needle-like characteristics start to form as it breaks down to PbI_2 (Leguy et al. 2015).

Laser beam-induced current imaging (LBIC) offers spatially resolved data on alterations in the local electrical environment by extending these measurements to whole devices. Due to the surface passivation of flaws, in situ tests reveal that the EQE somewhat rises after the device has been initially exposed to moisture. The feasibility of collecting holes at the spiro-MeOTAD junction then declines. As a result of increased exposure, hydrate phases develop, drastically lowering EQE. These findings demonstrate the connection between the actual functionality of a PSC and the causes of humidity-induced degradation discussed above.

2.2.2.3 Humidity-Assisted Ion Migration

It is evident from the preceding section that the $MAPbI_3$ matrix is highly hygroscopic and that prolonged exposure to high RH causes structural changes that are unfavorable for PSC operation. Yet, several investigations have since demonstrated that ion migration brought on by humidity might play an equally significant role in cell destruction. Fransishyn et al. conducted an operando GIWAXS investigation of MAPbI3-based PSCs in a humid condition. They could link perovskite structure alterations to device performance by obtaining GIWAXS patterns and I–V curves on the same sample. Their findings indicated that using silver electrodes caused performance to decline with increasing humidity; nevertheless, the GIWAXS data revealed no breakdown by-products, except a very slight reduction within the magnitude of the (110) perovskite peak (Fransishyn et al. 2018). Similar studies using gold electrodes showed a significant drop in the device performance, indicating the oxidative corrosion of the electrodes. These findings have been clarified by relying on ion diffusion permitted by moisture.

Moisture-induced ion migration also explains the peculiar behavior connected to the I–V curves of the FTO/ZnO/$MAPbI_3$/P3HT/Au devices. The peak close to

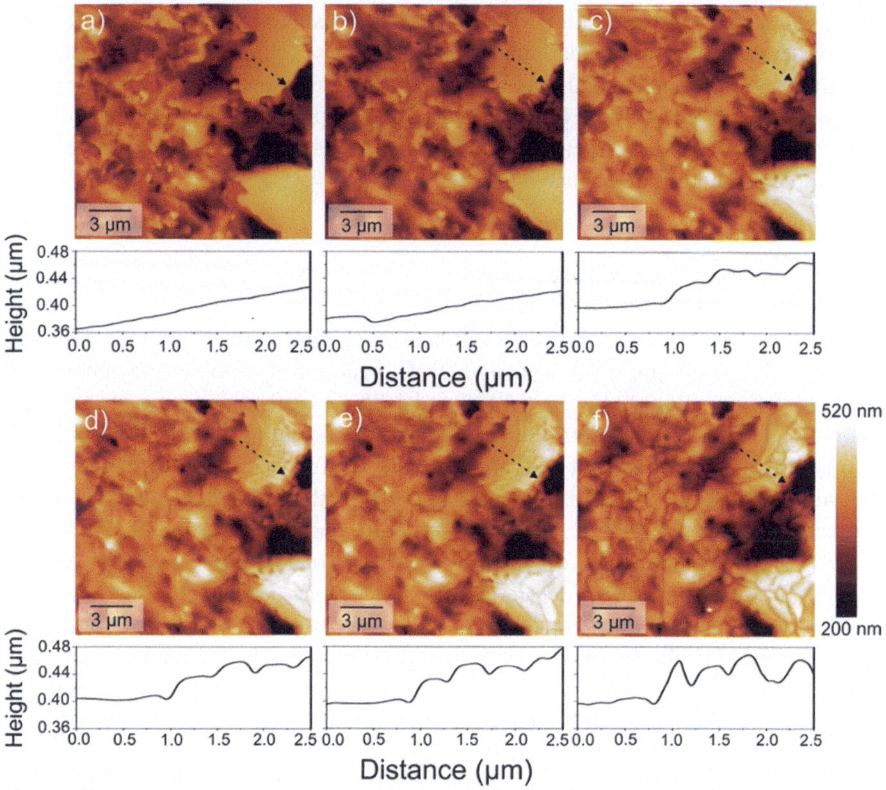

Fig. 2.11 SFM images and profiling of $MAPbI_3$ PSC under 80% RH, Reproduced from Ref. Li et al. (2016), Copyright (2016) with permission of the American Chemical Society

the maximum power point is a characteristic of cells with a high degree of hysteresis. This feature only becomes apparent once the device has been subjected to an RH of 85% rather than right away. This suggests that the hysteresis is caused by water absorption inside the device and an enhancement in the degree of ion movement. The degree of hysteresis has also increased with moisture absorption in other PSCs. Identical operando pXRD tests were conducted by Chen et al. on FTO/TiO_2/$MAPbI_3$/spiro-MeOTAD/Au devices at mild relative humidity (RH ~ 65% and ~ 90%) (Chen et al. 2017). The results showed a three-stage breakdown mechanism. The hydration of $MAPbI_3$ was responsible for the initial loss of fill factor. When PbI_2 appeared, the second stage displayed a decline in device performance. Peaks corresponding to PbIOH occurred at the final stage of degradation; now, the gadget is practically non-functional.

According to the authors, steady irradiation and moisture-induced degradation have a decomposition route. Because the PbIOH was only detected when spiro-MeOTAD was utilized, it was assumed that the HTL choice also substantially impacted the decomposition process. A portion of the lead is later reoxidized, creating

$PbO{\cdot}xH_2O$, as shown by in situ pXRD. The pXRD result confirmed that there is no PbI_2. Hence, it was assumed that this mechanism was specific for perovskite devices with aluminum electrodes and was an Al-induced breakdown mechanism. Ex situ XPS studies on thin aluminum films placed right on top of $MAPbI_3$ provide evidence in favor of these findings; a peak resembling metallic lead had been discovered, but the metallic aluminum had nearly vanished.

2.2.2.4 Thermal Degradation

Since solar modules must inevitably function in direct sunlight, their nominal working cell temperature is often around 50 °C. In addition, modules may encounter temperature fluctuations between − 40 °C and + 85 °C, depending on the surrounding environment. Hence, long-term thermal stability is essential for a PSC to pass the thermal cycling and moist heat tests stipulated by the International Electro-technical Commission (IEC 61215) (Holzhey et al. 2018). Nonetheless, a quick analysis shows how temperature affects both perovskite structure and its performance. At 56 °C, which is precisely within the operating range of a solar module, $MAPbI_3$ experiences a phase transition from tetragonal to cubic. Several investigations have also demonstrated that prolonged annealing of perovskite layers above 100 °C can cause them to disintegrate thermally. A thorough understanding of how temperature affects PSC deterioration is needed.

Early experiments showed that the visual color changed from dark brown to yellow over hours when $MAPbI_3$ films were thermally annealed at temperatures higher than 150 °C. This color change can be observed by ex situ UV–visible spectroscopy. The structural alterations that accompanied this transition were clarified by in situ GIWAXS measurements, which demonstrated that $MAPbI_3$ films break down to PbI_2 via intermediates. It was also discovered that if the heating period was long enough, temperatures as low as 80 °C were sufficient to start the breakdown (Eperon et al. 2014).

Although crystalline PbI_2 was unmistakably identified as one of the primary outcomes of heat decomposition by x-ray scattering techniques, it cannot deliver insights on volatile by-products. Nenon et al. employed temperature-programmed desorption mass spectrometry (TPD-MS) to clarify the degradation mechanism and linked the results with independent TGA studies (Nenon et al. 2016). While peaks corresponding to $CH_3NH_2^+$, CH_3^+, and NH_3^+ were found by their MS studies at the same temperature, its TGA data indicated a mass loss at 224 °C ± 3 °C. Juarez-Perez et al. discovered a distinct disintegration mechanism when $MAPbI_3$ was exposed to thermal stress in a helium flow. The gases generated during the TGA experiment could be immediately analyzed using thermal gravimetric differential thermal analysis (TG–DTA/MS), which is used in this study (Juarez-Perez et al. 2016). Two mass losses were visible on their TGA trace: one between 300 and 420 °C (related to the loss of MAI) and another at $T > 420$ °C (because of the loss of PbI_2; Fig. 2.12). When the gaseous species that evolved from the mass spectrometer are analyzed, it is discovered that the main by-products of MAI breakdown are NH_3 ($m/z = 17$)

and CH_3I ($m/z = 142$), which account for the majority of the signal. These thermal analysis results allow us to summarize how $MAPbI_3$ thermally decomposes:

$$MAPbI_3(s) \rightarrow HI(g) + CH_3NH_2(g) + PbI_2(s) \tag{2.3}$$

$$MAPbI_3(s) \rightarrow CH_3I(g) + NH_3(g) + PbI_2(s) \tag{2.4}$$

Simulations based on classical thermodynamic research indicate these two degradation routes (Ciccioli et al. 2018). Equation 2.3 is the kinetically preferred procedure since it requires very labile N–H and Pb–I bonds. In contrast, Eq. 2.4 is thermodynamically preferred as it is substantially slower because it requires breaking a less labile C–N bond. Additionally, the reaction between HI and CH_3NH_2 is reversible, whereas the breakdown of MA into CH_3I and NH_3 is irreversible. Equation 2.4 is only visible at extremely high temperatures due to the enormous activation energy. On the other hand, Eq. 2.3 is more pertinent to PSC deterioration because it can occur at temperatures as low as 76–144 °C.

Analyzing morphological and crystallographic changes can provide further information on thermal degradation. We may observe the structural and morphological alterations as they occur at the atomic level using in situ HR-TEM with SAED pattern. It is crucial to distinguish precisely between thermally induced and electron beam-induced breakdown since the high-intensity electron beam can significantly harm the area that is being targeted. According to an in situ TEM investigation by Fan et al., $MAPbI_3$ begins to break down at 85 °C under vacuum after 400 s. By rupturing the relatively loose Pb–I bonds, the degradation began to move in that direction. As a result, trigonal PbI_2 was formed, and gaseous CH_3NH_2 and HI were released (Fan et al. 2017). No intermediary phases were visible between the production of PbI_2 and the breakdown of $MAPbI_3$. In contrast, Kim et al. claim that $MAPbI_3$ thermally breaks down into PbI_2 through intermediate stages. Using an in situ HR-TEM, they could monitor the thermal breakdown procedure in real time while using the electron beam as a heat source. Fast Fourier transforms (FFTs) of the HR-TEM pictures and their specific information on the microstructural alterations in $MAPbI_3$ during heat degradation. The sample is heated at 20 °C/min from room temperature to 170 °C (Kim et al. 2018). The FFT image of the as-synthesized $MAPbI_3$ film revealed that it included many crystallographic domains, but no defects could be identified. Nevertheless, when the temperature had risen to 120 °C, a broad ring pattern representing an amorphous region began forming in the FFT image. The spot diffraction pattern showed that tetragonal PbI_2 started to emerge from the amorphous area above 130 °C. Moreover, ex situ SEM images display the morphological alterations at various temperatures. The grain boundaries of $MAPbI_3$ crystals merged with an enhancement in annealing temperature from 100 to 130 °C, whereas PbI_2 crystals were seen after annealing over 150 °C. A 30-min anneal at 170 °C produced a complete breakdown of PbI_2.

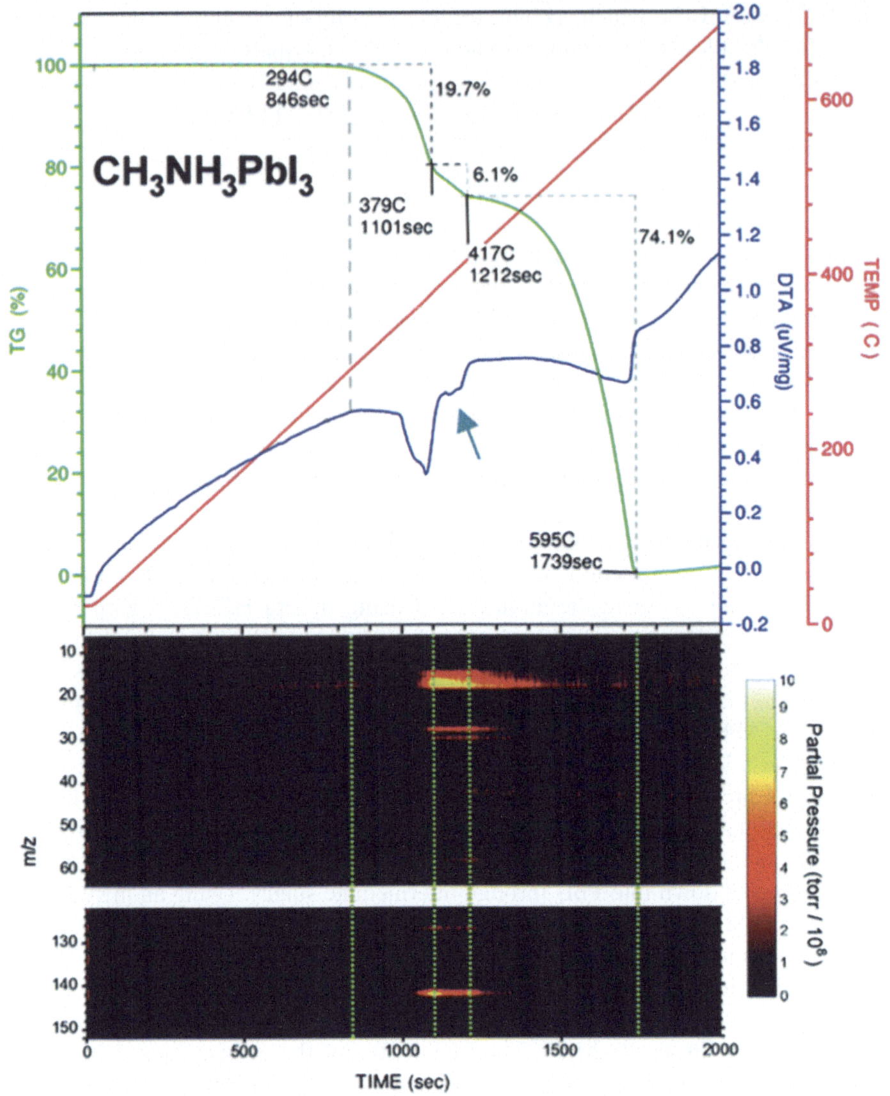

Fig. 2.12 Thermogravimetric analysis of the $MAPbI_3$ and *m*/*z* traces during the thermal degradation process, Reproduced from Ref. Juarez-Perez et al. (2016), Copyright (2016) with permission of the Royal Society of Chemistry

These thermal degradation investigations show that organic cations like MA^+ and FA^+ are unstable at higher temperatures. Hence, replacement inorganic cations, such as Cs, K, etc., may enhance the thermal stability of PSCs. In this regard, Zhang et al. demonstrated $CsPbBr_3$ nanocrystals, stable in a vacuum up to 417 °C, beyond which the nanocrystals sublimed. The thermal stability of $CsPbX_3$ (X = I, Br, or Cl) was explored by Liao et al. They discovered that the selection of the anion affects the breakdown and phase changes. According to in situ pXRD and in situ Raman spectroscopy, the thermal stability improved in the following sequence: $CsPbI_3$ is stable up to 250 °C, followed by $CsPbBr_3$ up to 450 °C, and $CsPbCl_3$ over 500 °C. Generally, $CsPbI_3$ breaks down into PbI_2 and CsI, while $CsPbBr_3$ breaks down into $PbBr_2$ and CsBr at high temperatures (Zhang et al. 2019).

2.2.2.5 Thermally Induced Ion Diffusion

It is evident from multiple investigations and theoretical predictions that fairly high temperatures (> 100 °C) can supply sufficient thermal energy to pass the activation barriers of the chemical reactions which result in perovskite disintegration. Enhanced ionic mobility inside the lattice can result in additional undesirable side effects in addition to perovskite disintegration. Even in the lack of substantial perovskite degradation, this can significantly decrease performance for PSCs. This is especially noticeable for PSCs built using silver electrodes. According to Li et al., ex situ performance measurements of PSCs with an FTO/NiO/$MAPbI_3$/PCBM/Ag structure at 85 °C revealed a considerable decline in performance. The PCBM/Ag contact had a build-up of I^-, I_2^- and CN^- ions, as revealed by the depth profiling. This suggests that MA and I^- ions are moving to the silver electrode through the PCBM layer. Iodide ions reacted with the silver electrode to form AgI, which was detected by pXRD data (Li et al. 2017). The outcomes unmistakably demonstrate how ion migration can cause device failure regardless of the absence of considerable perovskite disintegration.

Observations of thermally induced atom movement were made by Domanski et al. A substantial reduction in photocurrent was seen when PSCs with an FTO/*c*-TiO_2/*m*-TiO_2/$Cs_{0.05}(FA_{0.83}MA_{0.17})_{0.95}Pb(I_{0.83}Br_{0.17})_3$/spiro-MeOTAD/Au structure were subjected to thermal treatment for 15 h at 75 °C in an inert environment. Using ToF–SIMS, the authors discovered that the diffusion of contact Au across the spiro-MeOTAD HTL and penetration into the perovskite layer provide clear evidence for ion migration. The ensuing faults produced the reported drop in performance of the device (Domanski et al. 2016). Investigating FTO/*c*-TiO_2/*m*-TiO_2/$MAPbI_3$/spiro-MeOTAD/Au structured device using in situ STEM associated with EDX revealed additional proof of heat-activated ion migration (Fig. 2.13). Whereas EDX displays changes in elemental composition, the STEM displays morphological alterations with nano-scale spatial resolution. There aren't many changes once the sample is heated to 150 °C; however, when the temperature is raised to 175 °C, the perovskite layer begins to deteriorate while iodide seeps into the spiro-MeOTAD layer. At higher temperatures, the iodide diffusion is more prominent, and the perovskite layer exhibits morphological changes. At 225 °C, the deterioration is visible in both the HAADF and EDX maps, and at 250 °C, the $MAPbI_3$ had entirely disintegrated (Divitini et al. 2016).

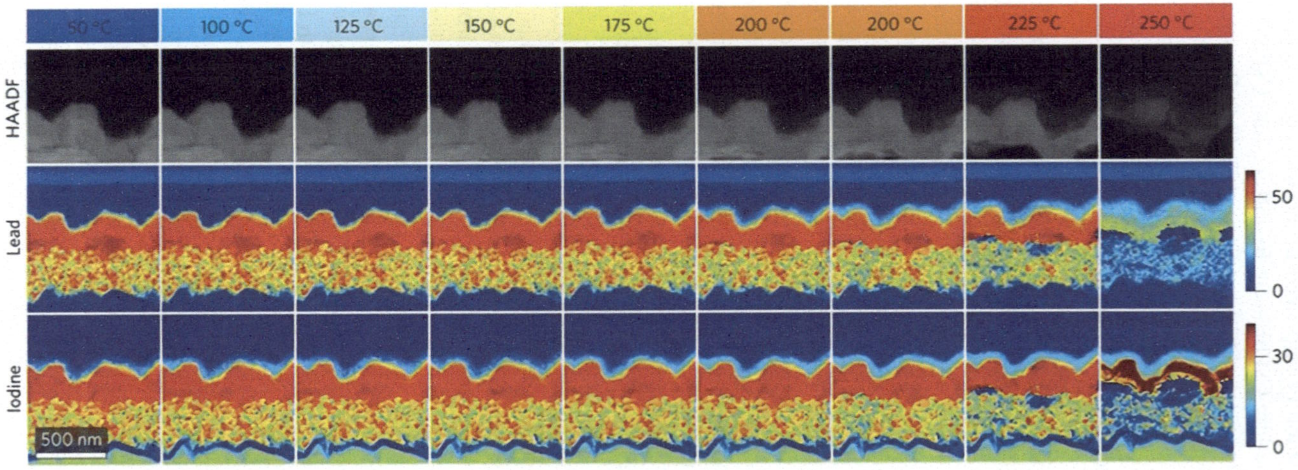

Fig. 2.13 Thermally induced ion migration confirmed via HAADF and EDX mappings. The phenomenon of iodine diffusing into the HTM is evident at low temperatures, whereas lead migration is initiated at elevated temperatures (~175 °C), Reproduced from Ref. Divitini et al. (2016), Copyright (2016) with permission of Springer Nature

2.2.2.6 Light-Induced Degradation

Light is the final significant environmental stressor that matters to PSCs. Although numerous investigations have now shown that encapsulated PSCs are stable under irradiation for hundreds of hours, a considerable amount of research still addresses photoinduced alterations to the PSC's structure and performance. One significant illustration is the light soaking technique, which is frequently discovered to improve device performance. Nevertheless, under prolonged exposure to irradiation, both reversible and permanent degradation have been seen. Several mechanisms, including the development of light-activated trap states, photoinduced ion segregation, photoinduced atom/ion mobility, and photodecomposition, have been proposed to explain this. Therefore, the advancement of this technology must have a more profound knowledge of the light-induced breakdown procedure (Nie et al. 2016).

The color of the $MAPbI_3$ films changed from black to gray when irradiated by white light in a vacuum, providing the first indication of photoinduced disintegration. When subjected to white light in an ultrahigh vacuum, comparable photobleaching effects are observed for $MAPbI_{3-x}Cl_x$ and $MAPbBr_{3-x}Cl_x$ films. These effects are intensified at high temperatures. Tang et al. investigated the structural changes brought on by light using in situ XRD to learn more about the photoinduced breakdown process. At 6 h of white light exposure in a vacuum, $MAPbI_3$ broke down into PbI_2 and metallic Pb, as shown in Fig. 2.14a, b, and after a longer exposure, PbI_2 continued to break down into Pb (Tang et al. 2016). Yang et al. investigations using $MAPb(I_{0.83}Br_{0.17})_3$, $FA_{0.83}MA_{0.17}Pb(I_{0.83}Br_{0.17})_3$, and $Cs_{0.1}(FA_{0.83}MA_{0.17})_{0.9}Pb(I_{0.83}Br_{0.17})_3$ also revealed the creation of PbI_2 and metallic Pb (Yang et al. 2018). After that, numerous groups have noted lead halides as a by-product of light-induced breakdown. X-ray scattering techniques must be further augmented by other techniques that are more susceptible to the organic component of the perovskite, as was the case with heat deterioration. In some studies, in situ XPS measurements conducted under open-circuit conditions revealed a generic mechanism for photoinduced breakdown.

$$MAPbI_3(s) \xrightarrow{\text{light}} \text{organic residue}(s) + NH_3(g) + HI(g) + PbI_2(s) \tag{2.5}$$

$$MAPbI_3(s) \xrightarrow{\text{light}} CH_3NH_2(g) + HI(g) + PbI_2(s) \tag{2.6}$$

$$PbI_2(s) \xrightarrow{\text{light}} Pb(s) + I_2(g) \tag{2.7}$$

Initially, solid PbI_2 and some gaseous components were formed due to the light-induced degradation of $MAPbI_3$ film, and then this solid PbI_2 disintegrates into metallic Pb and I_2 (Eq. 2.7). Mass spectrometry is a helpful tool for identifying the volatile by-products generated during the photoinduced breakdown. According to a mass spectrometry investigation by Song et al., $MAPbI_3$, CH_3NH_2, and HI are produced as the main breakdown products within 20 min of light irradiation (Eq. 2.6)

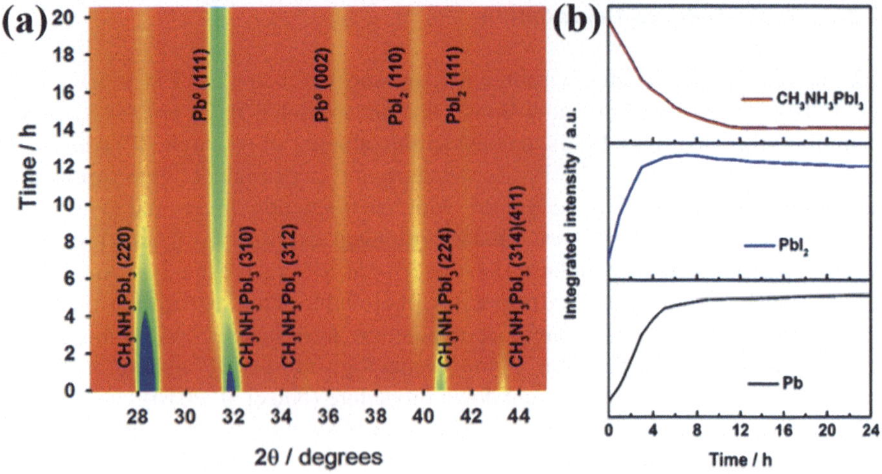

Fig. 2.14 **a** In situ XRD of $MAPbI_3$ PSCs at high temperature under the vacuum and illumination, and **b** estimated integral intensity using in situ XRD where the red peak is assigned to the $MAPbI_3$ and blue and Black peaks were designated to the PbI_2 and Pb respectively, Reproduced from Ref. Tang et al. (2016), Copyright (2016) with permission of the Royal Society of Chemistry

(Song et al. 2018b). Moreover, the presence of H_2, NH_3, and I_2 was noted. This may result from photodecomposition or the dissociation of CH_3NH_2 and HI caused by electron assault during the ionization process (Eq. 2.5). Numerous investigations have demonstrated that light-induced decomposition proceeds more quickly in air than in a vacuum or an inert atmosphere. Within a few hours, $MAPbI_3$ films transform from black to translucent or yellow when stored in the air under constant illumination. This change in optical characteristics is also seen in ex situ UV–visible investigations. In situ, XRD investigation revealed that the $MAPbI_3$ breaks down into PbI_2 in less than an hour when exposed to light and air. It breaks down even more on extended exposures into lead salts such as PbO, $PbCO_3$, and $Pb(OH)_2$.

There have been many proposed breakdown paths in the air under light irradiation, but the most widely recognized one can be expressed as follows:

$$MAPbI_3 \xrightarrow{h\nu \geq E_g} MAPbI_3^* \tag{2.8}$$

$$MAPbI_3^* + O_2 \rightarrow MAPbI_3^+ + O_2^{\cdot-} \tag{2.9}$$

$$MAPbI_3 + O_2^{\cdot-} \rightarrow 4CH_3NH_2 + 4PbI_2 + 2I_2 + 2H_2O \tag{2.10}$$

The formation of electron and hole pairs results from the photoexcitation of $MAPbI_3$. A volatile superoxide anion is created when a photoelectron moves from the $MAPbI_3$ conduction band to the molecular oxygen. This superoxide anion then targets the perovskite, which leads to the deprotonation of the MA cation and the oxidation of iodide. As demonstrated by Bryant et al., when devices were consistently exposed to air, their performance suffered significantly. For $TiO_2/MAPbI_3$ films, similar degradation paths were seen (Bryant et al. 2016).

2.3 Conclusion

The necessity of utilizing the plentiful solar resources on Earth has been brought to light by the effort to produce electricity from renewable energy sources. Research has concentrated on increasing solar cell efficiency and lowering manufacturing costs as solar PV capacity continues to rise globally. Currently under development, third-generation PSCs have the potential to rule the solar PV market eventually. PSCs were found to be the most promising candidate, with a PCE of 25.8% reaching in less than ten years.

Lead halide perovskites have achieved record PCEs. However, they are frequently lacking in their long-term stability due to their fragile bonds and mobile ions. Our understanding of the chemical and physical alterations that moisture, heat, and light may all cause in the perovskite lattice has dramatically improved due to advanced characterization techniques and both in situ and operando methodologies. The systematic abolition of these degrading pathways has already involved much labor. Further advancements are hoped to result in a commercially viable and fiercely competitive photovoltaic technology because there are currently numerous PSCs with good performance and respectably long lifetimes.

References

Arora N, Dar MI, Hinderhofer A et al (2017) Perovskite solar cells with CuSCN hole extraction layers yield stabilized efficiencies greater than 20%. Science (80-) 358:768–771. https://doi.org/10.1126/science.aam5655

Askar AM, Bernard GM, Wiltshire B et al (2017) Multinuclear magnetic resonance tracking of hydro, thermal, and hydrothermal decomposition of $CH_3NH_3PbI_3$. J Phys Chem C 121:1013–1024. https://doi.org/10.1021/acs.jpcc.6b10865

Bai S, Da P, Li C et al (2019) Planar perovskite solar cells with long-term stability using ionic liquid additives. Nature 571:245–250. https://doi.org/10.1038/s41586-019-1357-2

Bertoluzzi L, Boyd CC, Rolston N et al (2020) Mobile ion concentration measurement and open-access band diagram simulation platform for halide perovskite solar cells. Joule 4:109–127. https://doi.org/10.1016/j.joule.2019.10.003

Bi E, Tang W, Chen H et al (2019) Efficient perovskite solar cell modules with high stability enabled by iodide diffusion barriers. Joule 3:2748–2760. https://doi.org/10.1016/j.joule.2019.07.030

Bryant D, Aristidou N, Pont S et al (2016) Light and oxygen induced degradation limits the operational stability of methylammonium lead triiodide perovskite solar cells. Energy Environ Sci 9:1655–1660. https://doi.org/10.1039/c6ee00409a

Bush KA, Bailie CD, Chen Y et al (2016) Thermal and environmental stability of semi-transparent perovskite solar cells for tandems enabled by a solution-processed nanoparticle buffer layer and sputtered ITO electrode. Adv Mater 28:3937–3943. https://doi.org/10.1002/adma.201505279

Chen BA, Lin JT, Suen NT et al (2017) In situ identification of photo- and moisture-dependent phase evolution of perovskite solar cells. ACS Energy Lett 2:342–348. https://doi.org/10.1021/acsenergylett.6b00698

Cheng Y, Li HW, Qing J et al (2016) The detrimental effect of excess mobile ions in planar $CH_3NH_3PbI_3$ perovskite solar cells. J Mater Chem A 4:12748–12755. https://doi.org/10.1039/c6ta05053k

Cheng Y, Ding L et al (2021) Pushing commercialization of perovskite solar cells by improving their intrinsic stability. Energy Environ Sci 14:3233–3255. https://doi.org/10.1039/d1ee00493j

Chowdhury MS, Rahman KS, Chowdhury T et al (2020) An overview of solar photovoltaic panels' end-of-life material recycling. Energy Strateg Rev 27:100431. https://doi.org/10.1016/j.esr.2019.100431

Ciccioli A, Latini A et al (2018) Thermodynamics and the intrinsic stability of lead halide perovskites $CH_3NH_3PbX_3$. J Phys Chem Lett 9:3756–3765. https://doi.org/10.1021/acs.jpclett.8b00463

Dedecker K, Grancini G (2020) Dealing with lead in hybrid perovskite: a challenge to tackle for a bright future of this technology? Adv Energy Mater 10:1–8. https://doi.org/10.1002/aenm.202001471

Divitini G, Cacovich S, Matteocci F et al (2016) In situ observation of heat-induced degradation of perovskite solar cells. Nat Energy 1. https://doi.org/10.1038/NENERGY.2015.12

Domanski K, Correa-Baena JP, Mine N et al (2016) Not all that glitters is gold: metal-migration-induced degradation in perovskite solar cells. ACS Nano 10:6306–6314. https://doi.org/10.1021/acsnano.6b02613

Eperon GE, Stranks SD, Menelaou C et al (2014) Formamidinium lead trihalide: a broadly tunable perovskite for efficient planar heterojunction solar cells. Energy Environ Sci 7:982–988. https://doi.org/10.1039/c3ee43822h

Fan Z, Xiao H, Wang Y et al (2017) Layer-by-layer degradation of methylammonium lead tri-iodide perovskite microplates. Joule 1:548–562. https://doi.org/10.1016/j.joule.2017.08.005

Fransishyn KM, Kundu S, Kelly TL (2018) Elucidating the failure mechanisms of perovskite solar cells in humid environments using in situ grazing-incidence wide-angle x-ray scattering. ACS Energy Lett 3:2127–2133. https://doi.org/10.1021/acsenergylett.8b01300

Galagan Y (2020) Stability of perovskite PV modules. JPhys Energy 2. https://doi.org/10.1088/2515-7655/ab7077

Guo S, Sun X, Ding C et al (2020) Non-uniform chemical corrosion of metal electrode of p–i–n type of perovskite solar cells caused by the diffusion of CH_3NH_3I. Energy Technol 8:1–11. https://doi.org/10.1002/ente.202000250

Holzhey P, Saliba M et al (2018) A full overview of international standards assessing the long-term stability of perovskite solar cells. J Mater Chem A 6:21794–21808. https://doi.org/10.1039/C8TA06950F

Hou Y, Du X, Scheiner S et al (2017) A generic interface to reduce the efficiency-stability-cost gap of perovskite solar cells. Science (80-) 358:1192–1197. https://doi.org/10.1126/science.aao5561

Jacobsson TJ, Schwan LJ, Ottosson M et al (2015) Determination of thermal expansion coefficients and locating the temperature-induced phase transition in methylammonium lead perovskites using x-ray diffraction. Inorg Chem 54:10678–10685. https://doi.org/10.1021/acs.inorgchem.5b01481

Jiang Y, Qiu L, Juarez-Perez EJ et al (2019) Reduction of lead leakage from damaged lead halide perovskite solar modules using self-healing polymer-based encapsulation. Nat Energy 4:585–593. https://doi.org/10.1038/s41560-019-0406-2

Juarez-Perez EJ, Hawash Z, Raga SR et al (2016) Thermal degradation of $CH_3NH_3PbI_3$ perovskite into NH_3 and CH_3I gases observed by coupled thermogravimetry-mass spectrometry analysis. Energy Environ Sci 9:3406–3410. https://doi.org/10.1039/c6ee02016j

Kato Y, Ono LK, Lee MV et al (2015) Silver iodide formation in methyl ammonium lead iodide perovskite solar cells with silver top electrodes. Adv Mater Interfaces 2:2–7. https://doi.org/10.1002/admi.201500195

Kerner RA, Schulz P, Christians JA et al (2019) Reactions at noble metal contacts with methylammonium lead triiodide perovskites: role of underpotential deposition and electrochemistry. APL Mater 7. https://doi.org/10.1063/1.5083812

Kim GW, Kang G, Malekshahi Byranvand M et al (2017) Gradated mixed hole transport layer in a perovskite solar cell: improving moisture stability and efficiency. ACS Appl Mater Interfaces 9:27720–27726. https://doi.org/10.1021/acsami.7b07071

Kim TW, Shibayama N, Cojocaru L et al (2018) Real-time in situ observation of microstructural change in organometal halide perovskite induced by thermal degradation. Adv Funct Mater 28. https://doi.org/10.1002/adfm.201804039

Leguy AMA, Hu Y, Campoy-Quiles M et al (2015) Reversible hydration of $CH_3NH_3PbI_3$ in films, single crystals, and solar cells. Chem Mater 27:3397–3407. https://doi.org/10.1021/acs.chemmater.5b00660

Leijtens T, Eperon GE, Pathak S et al (2013) Overcoming ultraviolet light instability of sensitized TiO_2 with meso-superstructured organometal tri-halide perovskite solar cells. Nat Commun 4:1–8. https://doi.org/10.1038/ncomms3885

Li D, Bretschneider SA, Bergmann VW et al (2016) Humidity-induced grain boundaries in $MAPbI_3$ perovskite films. J Phys Chem C 120:6363–6368. https://doi.org/10.1021/acs.jpcc.6b00335

Li J, Dong Q, Li N, Wang L (2017) Direct evidence of ion diffusion for the silver-electrode-induced thermal degradation of inverted perovskite solar cells. Adv Energy Mater 7:1–8. https://doi.org/10.1002/aenm.201602922

Li X, Zhang F, He H et al (2020) On-device lead sequestration for perovskite solar cells. Nature 578:555–558. https://doi.org/10.1038/s41586-020-2001-x

Liang L, Cai Y, Li X et al (2018) All that glitters is not gold: Recent progress of alternative counter electrodes for perovskite solar cells. Nano Energy 52:211–238. https://doi.org/10.1016/j.nanoen.2018.07.049

Liu M, Chen Y, Tan CS et al (2019) Lattice anchoring stabilizes solution-processed semiconductors. Nature 570:96–101. https://doi.org/10.1038/s41586-019-1239-7

Mei A, Li X, Liu L et al (2014) A hole-conductor-free, fully printable mesoscopic perovskite solar cell with high stability. Science (80-) 345:295–298. https://doi.org/10.1126/science.1254763

Meng L, You J, Yang Y (2018) Addressing the stability issue of perovskite solar cells for commercial applications. Nat Commun 9:1–4. https://doi.org/10.1038/s41467-018-07255-1

Meng X, Xing Z, Hu X et al (2020) Stretchable perovskite solar cells with recoverable performance. Angew Chemie—Int Ed 59:16602–16608. https://doi.org/10.1002/anie.202003813

Ming W, Yang D, Li T et al (2018) Formation and diffusion of metal impurities in perovskite solar cell material $CH_3NH_3PbI_3$: implications on solar cell degradation and choice of electrode. Adv Sci 5. https://doi.org/10.1002/advs.201700662

Nenon DP, Christians JA, Wheeler LM et al (2016) Structural and chemical evolution of methylammonium lead halide perovskites during thermal processing from solution. Energy Environ Sci 9:2072–2082. https://doi.org/10.1039/c6ee01047d

Nie W, Blancon JC, Neukirch AJ et al (2016) Light-activated photocurrent degradation and self-healing in perovskite solar cells. Nat Commun 7:1–9. https://doi.org/10.1038/ncomms11574

NREL (2023) Best research-cell efficiency chart, Best Research-Cell Efficiency Chart | Photovoltaic Research | NREL

Park NG (2015) Perovskite solar cells: an emerging photovoltaic technology. Mater Today 18:65–72. https://doi.org/10.1016/j.mattod.2014.07.007

Rolston N, Bush KA, Printz AD et al (2018) Engineering stress in perovskite solar cells to improve stability. Adv Energy Mater 8:1–7. https://doi.org/10.1002/aenm.201802139

Saliba M et al (2016a) Incorporation of rubidium cations into perovskite solar cells improves photovoltaic performance. Science (80-) 354:206–209. https://doi.org/10.1126/science.aah55

Saliba M, Matsui T, Seo JY et al (2016b) Cesium-containing triple cation perovskite solar cells: Improved stability, reproducibility and high efficiency. Energy Environ Sci 9:1989–1997. https://doi.org/10.1039/c5ee03874j

Saris S, Niemann V, Mantella V et al (2019) Understanding the mechanism of metal-induced degradation in perovskite nanocrystals. Nanoscale 11:19543–19550. https://doi.org/10.1039/c9nr06328e

Schlipf J, Bießmann L, Oesinghaus L et al (2018) In situ monitoring the uptake of moisture into hybrid perovskite thin films. J Phys Chem Lett 9:2015–2021. https://doi.org/10.1021/acs.jpclett.8b00687

Shlenskaya NN, Belich NA, Grätzel M et al (2018) Light-induced reactivity of gold and hybrid perovskite as a new possible degradation mechanism in perovskite solar cells. J Mater Chem A 6:1780–1786. https://doi.org/10.1039/c7ta10217h

Song Z, Shrestha N, Watthage SC et al (2018a) Impact of moisture on photoexcited charge carrier dynamics in methylammonium lead halide perovskites. J Phys Chem Lett 9:6312–6320. https://doi.org/10.1021/acs.jpclett.8b02595

Song Z, Wang C, Phillips AB et al (2018b) Probing the origins of photodegradation in organic-inorganic metal halide perovskites with time-resolved mass spectrometry. Sustain Energy Fuels 2:2460–2467. https://doi.org/10.1039/c8se00358k

Tang X, Brandl M, May B et al (2016) Photoinduced degradation of methylammonium lead triiodide perovskite semiconductors. J Mater Chem A 4:15896–15903. https://doi.org/10.1039/c6ta06497c

Turren-Cruz SH, Hagfeldt A, Saliba M (2018) Methylammonium-free, high-performance, and stable perovskite solar cells on a planar architecture. Science (80-) 362:449–453. https://doi.org/10.1126/science.aat3583

Unger EL, Hoke ET, Bailie CD et al (2014) Hysteresis and transient behavior in current-voltage measurements of hybrid-perovskite absorber solar cells. Energy Environ Sci 7:3690–3698. https://doi.org/10.1039/c4ee02465f

Van Reenen S, Kemerink M, Snaith HJ (2015) Modeling anomalous hysteresis in perovskite solar cells. J Phys Chem Lett 6:3808–3814. https://doi.org/10.1021/acs.jpclett.5b01645

Wang Q, Chen B, Liu Y et al (2017) Scaling behavior of moisture-induced grain degradation in polycrystalline hybrid perovskite thin films. Energy Environ Sci 10:516–522. https://doi.org/10.1039/c6ee02941h

Wang P, Zhang X, Zhou Y et al (2018) Solvent-controlled growth of inorganic perovskite films in dry environment for efficient and stable solar cells. Nat Commun 9:1–7. https://doi.org/10.1038/s41467-018-04636-4

Wang L, Zhou H, Hu J et al (2019a) A Eu^{3+}-Eu^{2+} ion redox shuttle imparts operational durability to Pb-I perovskite solar cells. Science (80-) 363:265–270. https://doi.org/10.1126/science.aau5701

Wang Y, Wu T, Barbaud J et al (2019b) Stabilizing heterostructures of soft perovskite semiconductors. Science (80-) 365:687–691. https://doi.org/10.1126/science.aax8018

Werner J, Boyd CC, Moot T et al (2020) Learning from existing photovoltaic technologies to identify alternative perovskite module designs. Energy Environ Sci 13:3393–3403. https://doi.org/10.1039/d0ee01923b

Wu WQ, Rudd PN, Ni Z et al (2020) Reducing surface halide deficiency for efficient and stable iodide-based perovskite solar cells. J Am Chem Soc 142:3989–3996. https://doi.org/10.1021/jacs.9b13418

Xu J, Dong H, Xi J et al (2020a) Local nearly non-strained perovskite lattice approaching a broad environmental stability window of efficient solar cells. Nano Energy 75:104940. https://doi.org/10.1016/j.nanoen.2020.104940

Xu S, Liu G, Zheng H et al (2020b) Boosting photovoltaic performance and stability of super-halogen-substituted perovskite solar cells by simultaneous methylammonium immobilization and vacancy compensation. ACS Appl Mater Interfaces 12:8249–8259. https://doi.org/10.1021/acsami.9b21074

Yang J, Siempelkamp BD, Liu D, Kelly TL (2015) Investigation of $CH_3NH_3PbI_3$ degradation rates and mechanisms in controlled humidity environments using in situ techniques. ACS Nano 9:1955–1963. https://doi.org/10.1021/nn506864k

Yang J, Hong Q, Yuan Z et al (2018) Unraveling photostability of mixed cation perovskite films in extreme environment. Adv Opt Mater 6. https://doi.org/10.1002/adom.201800262

Yang S, Chen S, Mosconi E et al (2019) Stabilizing halide perovskite surfaces for solar cell operation with wide-bandgap lead oxysalts. Science (80-) 365:473–478. https://doi.org/10.1126/science.aax3294

Yuan Y, Huang J (2016) Ion Migration in organometal trihalide perovskite and its impact on photovoltaic efficiency and stability. Acc Chem Res 49:286–293. https://doi.org/10.1021/acs.accounts.5b00420

Zhang T, Meng X, Bai Y et al (2017) Profiling the organic cation-dependent degradation of organolead halide perovskite solar cells. J Mater Chem A 5:1103–1111. https://doi.org/10.1039/C6TA09687E

Zhang H, Ren X, Chen X et al (2018) Improving the stability and performance of perovskite solar cells: via off-the-shelf post-device ligand treatment. Energy Environ Sci 11:2253–2262. https://doi.org/10.1039/c8ee00580j

Zhang C, Fernando JFS, Firestein KL et al (2019) Thermal stability of CsPbBr3 perovskite as revealed by in situ transmission electron microscopy. APL Mater 7:1–7. https://doi.org/10.1063/1.5108849

Zhao Y, Liang C, Zhang H et al (2015) Anomalously large interface charge in polarity-switchable photovoltaic devices: an indication of mobile ions in organic-inorganic halide perovskites. Energy Environ Sci 8:1256–1260. https://doi.org/10.1039/c4ee04064c

Zhao J, Cai B, Luo Z et al (2016a) Investigation of the hydrolysis of perovskite organometallic halide $CH_3NH_3PbI_3$ in humidity environment. Sci Rep 6:1–6. https://doi.org/10.1038/srep21976

Zhao J, Zheng X, Deng Y et al (2016b) Is Cu a stable electrode material in hybrid perovskite solar cells for a 30-year lifetime? Energy Environ Sci 9:3650–3656. https://doi.org/10.1039/c6ee02980a

Zhao J, Deng Y, Wei H et al (2017) Strained hybrid perovskite thin films and their impact on the intrinsic stability of perovskite solar cells. Sci Adv 3. https://doi.org/10.1126/sciadv.aao5616

Zhou Y, Yang M, Vasiliev AL et al (2015) Growth control of compact $CH_3NH_3PbI_3$ thin films via enhanced solid-state precursor reaction for efficient planar perovskite solar cells. J Mater Chem A 3:9249–9256. https://doi.org/10.1039/c4ta07036d

Zhu C, Niu X, Fu Y et al (2019) Strain engineering in perovskite solar cells and its impacts on carrier dynamics. Nat Commun 10. https://doi.org/10.1038/s41467-019-08507-4

Chapter 3
Operational Stability Management Approaches for Perovskite Solar Cells

Abstract Perovskite solar cells (PSCs) rapid progress is driven by their exceptional optoelectronic properties. Transitioning to commercialization resolves efficiency and cost concerns, focusing on addressing the critical challenge of a limited operational lifetime under environmental aging. Modified International Electrotechnical Commission (IEC) testing standards provide a robust foundation for assessing PSC stability. Some PSCs pass tests with enhanced films, advanced architectures, and effective stabilization methods. The ionic and soft nature of perovskites introduces challenges like ion transportation and phase separation. Developing functionally steady PSCs is crucial for commercializing perovskite PV technologies. Standardizing testing processes and reporting for stability results is essential, requiring detailed experimental procedures. The research community's intensified focus anticipates rapid improvements in operational stability, mirroring progress in PSC efficiency and facilitating commercialization.

Keywords Operational stability · IEC standards · Stability engineering · Additive engineering · Defect passivation · Morphological engineering · Cation–anion engineering · ISOS stability protocols

3.1 Introduction

Metal halide perovskite solar cells (MH-PSCs) received considerable attention as a viable photovoltaic technology because of their excellent optoelectronic properties, including high photoabsorption, prominent charge carrier mobility, and cost-effective manufacturing. In past ten years, the performance of the PSCs increased dramatically from 3.8% in 2009 to above 26% in 2024, surpassing the efficiency of other commercialized thin film solar cells and approaching the state-of-the-art efficiency of crystalline-silicon solar cells. This significant progress in efficiency is attributed to the excellent optoelectronic characteristics of the perovskite absorber and the utilization of expertise gained from extensive research and advancements in PV industry. Additionally, PSC production costs are predicted to be half as high as

A. Srivastava and P. M. Shirage, *Perovskite Solar Cells*, Springer Tracts in Electrical and Electronics Engineering, https://doi.org/10.1007/978-981-97-7624-5_3

those of crystalline-silicon solar cells, making PSCs an attractive option for large-scale commercialization (Li et al. 2019). Furthermore, by mounting perovskite layers on top of currently existing commercial solar cells, the unit power output can be enhanced, reducing the levelized cost of electricity. Despite these advantages, the lifetime of PSCs remains a significant hurdle for their commercialization, with early reports indicating a lifetime of only several seconds. While the lifetime of PSCs has improved significantly and currently stands at around one year under standard aging conditions, this is still considerably lower than the commercially available Si-solar cell's lifetime. As a result, the focus of research is now on enhancing the stability of the device and extending the PSC lifetime in real-world operating environments.

The reasons for the short lifespan of PSCs have yet to be fully understood, but there are intrinsic instability issues with perovskite semiconductors that make up these devices. Perovskite materials were generally attributed by the chemical formula ABX_3, where A, B, and X represent the monovalent cation, divalent metal ions, and halogens, respectively. Some of the most common A elements are like methylammonium (MA^+), formamidinium (FA^+), as well as Cs^+. Similarly, the most preferred B-site elements are Pb^{2+} and Sn^{2+} and X-sites are I^-, Br^-, and Cl^-. Any stoichiometry of these components in the form of ABX_3 can be incorporated into the perovskites crystal lattice (El-Mellouhi et al. 2016). However, unlike c-Si, which contains a highly stable atomic arrangement with a strong covalent bond, MHPs contain chemically vulnerable, and environmentally sensitive components. Due to their weak ionic and secondary bond interactions, these components are chemically susceptible to aging stimuli in the environment. The phase change, lattice collapse, chemical deterioration, and device failure are brought on by these stresses. Since the activation energy of ionic migration in perovskites is low, the ionic constituents are easily moved in response to external stimuli. This might result in phase separation or undesired ion redistribution, especially in mixed perovskites. Point defects including vacancies, interstitials, and anti-site defects frequently arise along with ion migration, and these can both help and worsen ion migration, which in turn causes the PSCs to deteriorate in different functional environments (Azpiroz et al. 2015).

The potential failure of the PSC device because of material instability is not limited to perovskite absorbers in photovoltaic devices. Charge transport layers, electrodes, and interfaces are also susceptible to degradation. Materials commonly used for electron and hole transport include TiO_2, SnO_2, PCBM, C_{60}, spiro-MeOTAD, PTAA, P3HT, NiO_x, and CuSCN. Long-term stability is crucial for PSCs to be commercially viable. Previous research has focused on the processes of PSC deterioration under diverse stressors and effective strategies to improve their stability. While early studies only investigated the shelf lifetime of PSCs, recent efforts have focused on improving their operational stability under realistic conditions. The longest reported lifespan of PSCs under operational environments is currently one year. However, commercial requirements exceed this lifespan, and a thorough analysis of recent advancements is urgently essential to comprehend and improve the operational stability of PSCs. It is necessary to construct a more structured overview that concentrates on the functional stability of perovskite devices within the context of their layered architecture, despite the fact that earlier reviews have examined the intrinsic and chemical stability of MH-PSCs (Park and Seok 2019).

The main goal of this chapter is to define operational stability clearly and discuss its importance in PSCs. The effective methods used to increase the operational robustness of PSCs are then compiled and examined, including changes made to the perovskite absorber itself, such as surface passivation, band-engineering, phase optimization, defect engineering, and morphological alterations. We also discuss the importance of material screening for charge transport layers and chemically stable electrodes. Encapsulation methods are briefly discussed as the best way to improve moisture stability. Furthermore, the review delves into the underlying mechanisms of these methods that can motivate the creation of more efficient approaches to fabricating the device. To promote progress in the community and create consensus on stability measurement techniques, a standardized operational stability parameter is suggested. This would enable different organizations to share useful information derived from standard testing results.

3.2 Requirement and Importance of Operational Stability

The functioning of solar cells involves converting luminous energy into charge carriers, such as free electrons and holes; these are subsequently captured and transferred to external circuits to produce electrical power. To be profitable, a PSC must be stable while operating a device under specific environmental conditions, particularly when exposed to light and electric load. This operational stability is critical for device behavior and lifetime, as it must withstand various operational stressors, including heat, humidity, and mechanical load. Unlike environmental stressors that can be prevented with external protection, operational stressors are inherent to working devices, necessitating improvements in the intrinsic resiliency of PSCs to ensure operational stability that meets commercial requirements. The lifespan of PSCs operating continuously under the exposure of 1-Sun illumination or near it while using an electric load, particularly in tracking the maximum power point, is the primary measure of operational stability and reflects the device's intrinsic stability. This parameter is crucial in assessing the feasibility of commercializing PSCs (Ma et al. 2020).

PSCs have lower lifetimes compared to silicon solar cells. This results from the organic–inorganic hybrid materials used in PSCs having constituents susceptible to the environment and a flexible ionic framework. This results in easy ion migration and significant alterations in perovskite films under sunlight, such as halide (Hoke Effect), cation segregation, and photodecomposition (Fig. 3.1). These changes occur within tens of minutes to hours under illumination, leading to various defects. The origin of photoinduced changes in perovskites has yet to be precisely discovered. Still, few studies suggest that increased halide vacancy concentration and ionic conductivity under sunlight exposure are in charge of phase separation and chemical breakdown. Additionally, carrier transport dynamics in perovskites are highly influenced by factors such as facet orientation and residual strain (Kim et al. 2018a; Brivio et al. 2016; Hoke et al. 2015).

The stability of PSCs is affected by various factors, including exposure to light, electric fields, and ion migration. Other types of charge transport layers (CTLs), like organic semiconductors and photocatalytic inorganic semiconductors, are also susceptible to photodegradation. When exposed to light, organic semiconductors can generate free radicals, causing cross-linking and increasing disorder within the semiconductor films. Light-induced oxygen desorption can create trap sites in photocatalytic inorganic CTLs such as TiO_2, leading to recombination and reduced charge transport. Furthermore, an electric field may promote ion migration, which affects p–n junctions' intrinsic electric field and prevents photogenerated carriers from passing

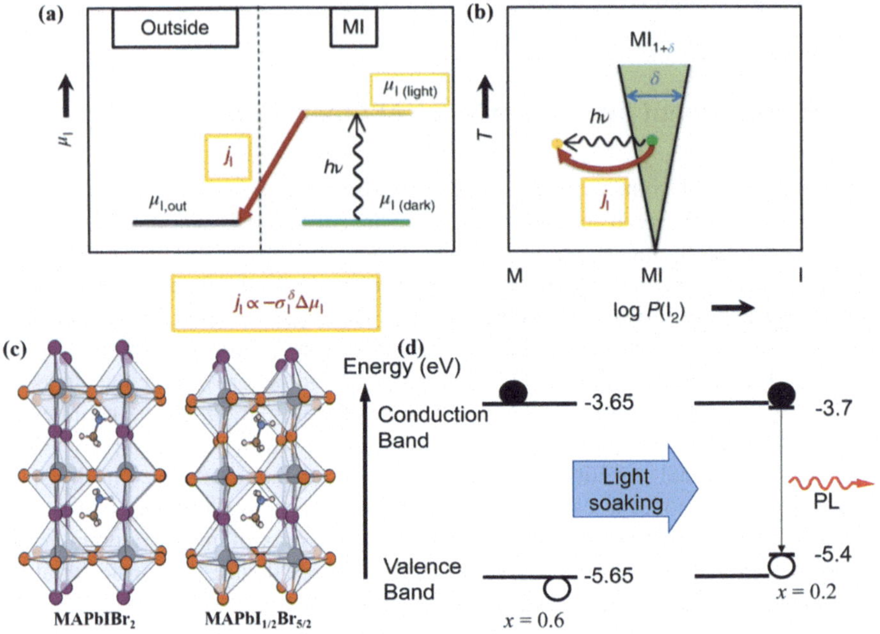

Fig. 3.1 **a, b** Schematics of photodecomposition pathway where I-deficient system originated falling outside the stability range, Reproduced from Ref. Kim et al. (2018a), Copyright (2018) with permission of Springer Nature. **c** Stable ordered $MAPbIBr_2$ and $MAPbI_{1/2}Br_{5/2}$ reducing the internal strain, Reproduced from Ref. Brivio et al. (2016), Copyright (2016) with permission of the American Chemical Society. **d** Schematic illustrating the proposed pathway by which photoinduced trap formation occurs through halide segregation, Reproduced from Ref. Hoke et al. (2015), Copyright (2015) with permission of the Royal Society of Chemistry

through. These factors pose significant challenges to the stability and commercialization of PSCs, and further research is needed to address these issues (Lin et al. 2018b).

3.3 IEC Operational Stability Standards

PSCs face challenges in commercialization due to their susceptibility to ambient environmental factors like heat, humidity, and mechanical load that can affect their stability. To examine the functional stability of various type of solar cells, the International Electro-technical Commission (IEC) has established testing standards, including the widely recognized IEC 61215 standard, which includes a series of stress tests that can bestow accelerated aging conditions to figure out the possible lifespan of a solar cell when used in real operation. Traditional PV technologies like c-Si and GaAs thin film solar cells have effectively passed this IEC operational stability test. However, PSCs must also meet this standard for commercialization. Companies like Frontier Energy Solutions and Saule Technologies are working to meet the IEC 61646 standard, which is now part of the IEC 61215. A prerequisite for producers and consumers to accept PSCs is passing the IEC 61215 (De Angelis et al. 2017). In 2018, Holzhey et al. carried out a thorough analysis of the current IEC 61215 tests, in which PSCs are required to remain stable under a range of circumstances, such as: 1-Sun illumination (1000 W/m^2); UV-light (15 kW h m^2 of UV irradiance from 280 to 400 nm); thermal cycling (40 to 85 °C); damp heat (85 °C and 85% RH); outdoor scenario (MPP at irradiance of minimum 60 kW h m^{-2}); humidity freeze (40 to 85 °C, 85% RH); hot-spot (50 $\pm$ 10 °C when used in 1000 W m^{-2} irradiance with the maximum power dissipation); hail impact (ice balls weighing 7.53 g and traveling at a speed of 82.8 km h^{-1}); and mechanical load (at least 2400 Pa), etc. shown in Fig. 3.2 (Holzhey et al. 2018). In addition to the tests mentioned in IEC 61215, high bias voltage should also be considered as an external electric field might cause ion transportation and solar cell deterioration, particularly for PSCs susceptible to minor electric fields. Extending the standards for PSC industrialization would be an excellent place to start, building on the IEC TS 62804-1 standard, which tests silicon solar cell stability under high bias voltages at high temperatures and humidity. Therefore, to achieve commercialization, PSCs must meet the necessary requirements, including passing established testing standards such as IEC 61215 and considering external stressors such as high bias voltage.

In order to assess PSCs' thermal stability, 200 cycles at varying temperatures between − 40 to 85 °C are performed in a thermal cycle test. This test is passed using an efficient polyisobutylene (PIB) as the moisture resistance in a low-temperature glass-glass encapsulation method. PIB may penetrate PSC materials without adversely affecting their performance, in contrast to other encapsulating techniques like EVA or UV-cured epoxy. This reduces the possibility of volatile decomposition products escaping from the PSCs during thermal investigation. Almost no power conversion efficiency (PCE) loss appears in the resultant PSC devices after

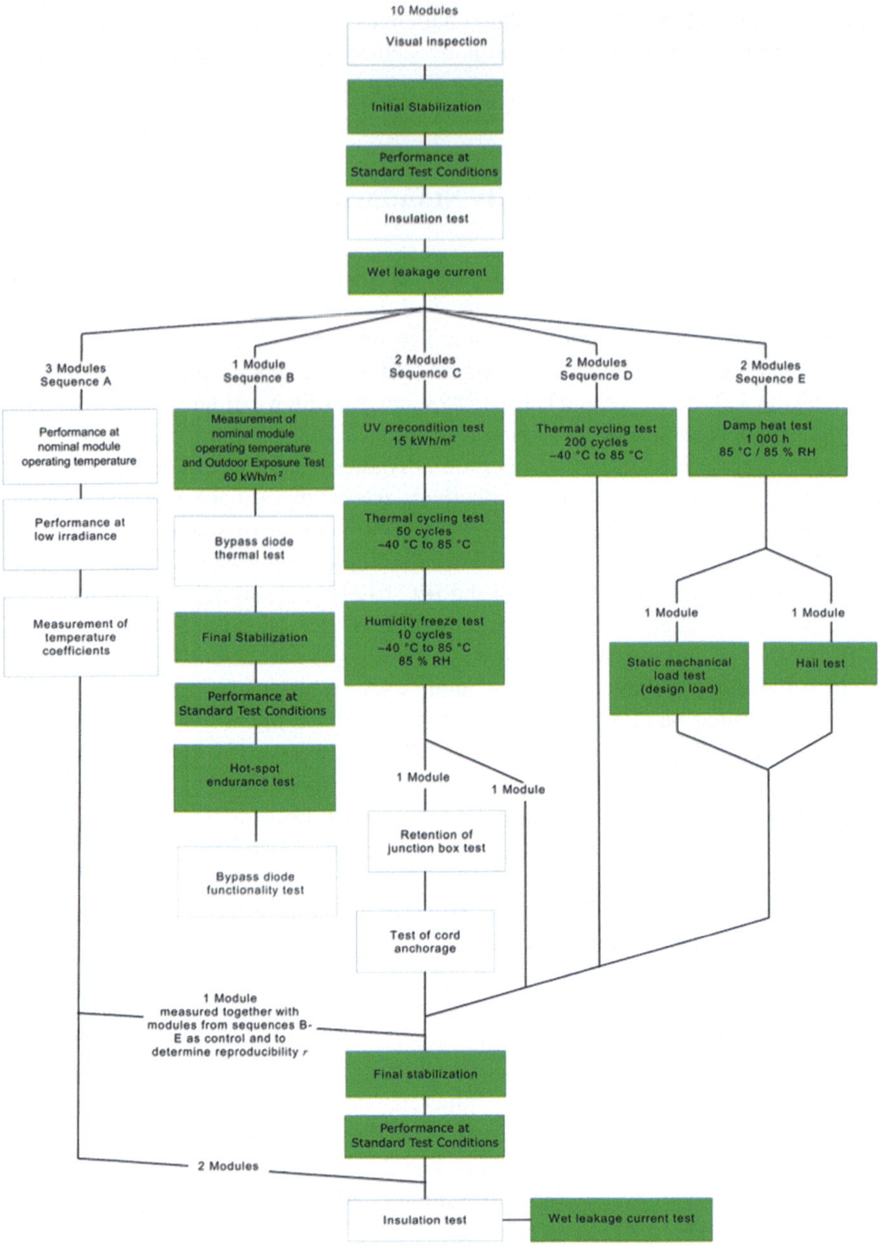

Fig. 3.2 Complete test flow for the design qualification and type approval of PV modules comprises five testing sequences: A, B, C, D, and E. Reproduced from Ref. IEC (2016), Copyright (2016) with permission of IEC Geneva, Switzerland http://www.iec.ch

200 heat cycles. The authors suggest that releasing volatile decomposition products is the main challenge to achieving thermal stability in PSCs and can be mitigated through efficient encapsulation. In conclusion, PSCs need to employ a thermally stable perovskite system and efficient encapsulation in order to pass the thermal IEC tests (Shi et al. 2017).

Bella et al. proposed using multi-functional photopolymers to improve the stability of PSCs under UV irradiation and outdoor conditions. They coated the PSC device's front side with a luminescent downshifting (LDS) fluoropolymer layer that converted UV radiation into visible light, thereby preventing potential damage of PSCs by UV radiation in the ambient illumination. In addition, a hydrophobic coating against outside moisture within the solar cells was created by developing a moisture-resistant fluoropolymer layer on the PSCs' backside. After aging tests, including UV irradiation in Ar and air (50% RH) over six months, the PSCs retained 98% of their initial PCE. Furthermore, after outdoor operation for three months, the PSCs maintained 95% of their initial efficiency. The effectiveness of the LDS material in enhancing the irradiation stability of PSCs was also supported by various studies (Bella et al. 2016).

Recent research has shown that encapsulation plays a critical role in enhancing the stability of PSCs and meeting various IEC requirements. Encapsulation materials are a barrier against moisture invasion and the leaking of volatile components involved in decomposition. Luminescent downshifting or organic therapeutic additives can also be used to functionalize the encapsulation materials and protect PSCs from a variety of external stimuli. For instance, thermos-cross-linking epoxy resin (ER) films applied to the upper sides of PSC modules are highly effective in protecting them from simulated hail impact, as shown in Fig. 3.3 (Jiang et al. 2019a). Such films exhibit natural healing properties and increased mechanical strength, enhancing PSCs' mechanical stability. Table 3.1 provides a summary of PSCs that have completed single IEC tests with the aid of encapsulation.

Achieving stable and reliable operation of photovoltaic devices under various stressors is a critical requirement for their commercialization. While numerous studies have focused on improving the PSC's stability within IEC testing standards, primarily within storage environments, there is a growing need to address the scientific challenges and technological limitations to enhance the operational stability of PSC devices. In this context, this text examines practical techniques to improve the functional stability of PSCs, providing valuable insights for future research aimed at developing PSCs that meet commercial demands.

3.4 Engineering on Perovskite Absorber Layer

In general, the unstable behavior of perovskite absorbers, while the functioning of PSCs, is primarily caused by the inherent composition of the materials, such as weak atomic bonding. This instability is also accelerated by defects in the lattice structures of the materials that can lead to additional deterioration in the crystal phases.

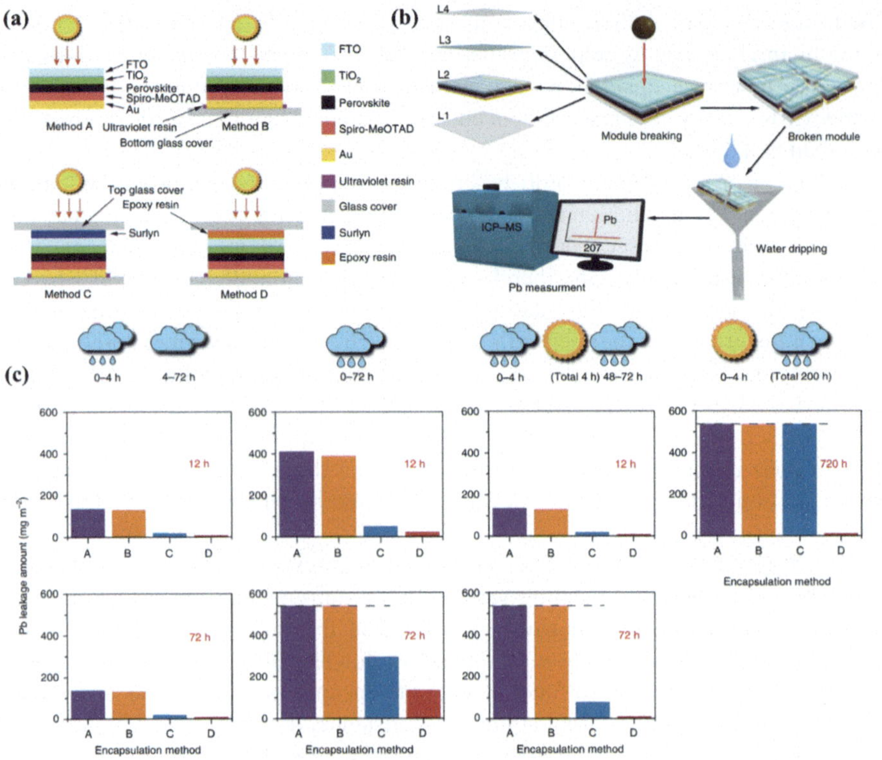

Fig. 3.3 **a**, **b** Schematic and experimental assessment of Pb-leakage from destructed PSC module, **c** representation of four-different weather conditions on Pb-leakage of encapsulated PSC modules. Reproduced from Ref. Jiang et al. (2019a), Copyright (2019) with permission of Springer Nature

Table 3.1 Some important IEC-61215 testing details for a standard PCE

Test	Perovskite	Testing parameters	PCE stability	References
Thermal cycle	$FAPbI_3$	Temp: − 40 to 85 °C dwell 15 min	200 thermal cycles	Shi et al. (2017)
Hail impact	$FA_{0.93}Cs_{0.07}PbI_3$	Metal ball dropping	Star-shaped cracks	Jiang et al. (2019b)
Damp heat	$FA_{1-x}Cs_xPbI_{3-y}Br_y$	Temp: 85 °C RH: 85%	1000 h	Bush et al. (2017)
UV irradiation	$FA_{1-x}MA_xPbI_{3-y}Br_y$	5 mA cm^{-2} in Ar 5 mA cm^{-2} in air RH: 50%	Six months (~ 98% PCE)	Bella et al. (2016)
Outdoor conditions	$FA_{1-x}MA_xPbI_{3-y}Br_y$	On the terrace	Three months (~95% PCE)	

Chemical Instability: The intrinsic compositions of halide perovskites, which consist of A-, B-, and X-site elements, are mainly responsible for their instability during PSC operation. These constituents are low-valence ions having large radius, resulting in relatively small ionic potential and comparably low lattice energy in the crystal (Fig. 3.4a, b) (Cappel et al. 2017). This allows the easy formation of crystals at low temperatures and makes the soft crystal lattices susceptible to different aging stressors like moisture, oxygen, heat, and radiance. Moreover, the chemical reactivity of organic ammonium ions at the A-site, which is improved by the polar surroundings of C–N or N–H bonding, can cause the decomposition of halide perovskites at high temperatures and their reaction with surrounding molecules (Kim et al. 2016). This chemical instability is mainly attributed to the interaction between the relevant constituents.

Existing defects: Polycrystalline perovskite films have a variety of defects due to their low formation energy, which can serve as a non-radiative trap center for free charge carriers, degrading the effectiveness of the device. Furthermore, the presence of grain boundaries and pinholes in these films can provide additional locations for defects and accelerate ion migration. This ion migration has the potential to cause further defects in perovskites and self-accelerated component losses, which will have a substantial effect on how carriers behave in perovskites when illuminated (Fig. 3.4c, d). Defects and ion migration are critical factors contributing to the operational instability of PSCs (Eames et al. 2015).

Structural and Phase instability: In perovskite materials that are active in light absorption, the crystal structure typically adopts a cubic phase, where [BX_6] octahedra form a framework at the corners of the cube, and the cubes center was occupied by A-site entities. However, residual strain brought on by mechanical, thermal, or light stresses as well as the Jahn–Teller effect can deform the real lattice structure (Fig. 3.4e, f) (Gui et al. 2018). Suppose the A-site ions are not compatible in size. Subsequently, volatile cubic frameworks exhibiting an unfavorable tolerance factor can be introduced, resulting in a photoinactive phase transition from the α- to the δ-phase. Furthermore, in cation or anion hybrid perovskites, lattice mismatch can result in phase separation under light. These phase variations impact PSCs' operational stability, atomic interaction, and lattice occupancy (Rolston et al. 2020).

Several techniques have been proposed to address the inherent instability of perovskites, such as incorporating functional additives, modifying the composition of perovskite materials, and creating a protective layer on the perovskite surface to address the inherent instability of perovskite. The subsequent discussion will delve deeper into these strategies.

3.4.1 Additive Engineering

Perovskite films are typically produced by solution processing that can introduce various defects in the material's composition and structure. Point defects, such as electron acceptors and electron donors, are commonly observed in perovskites. The

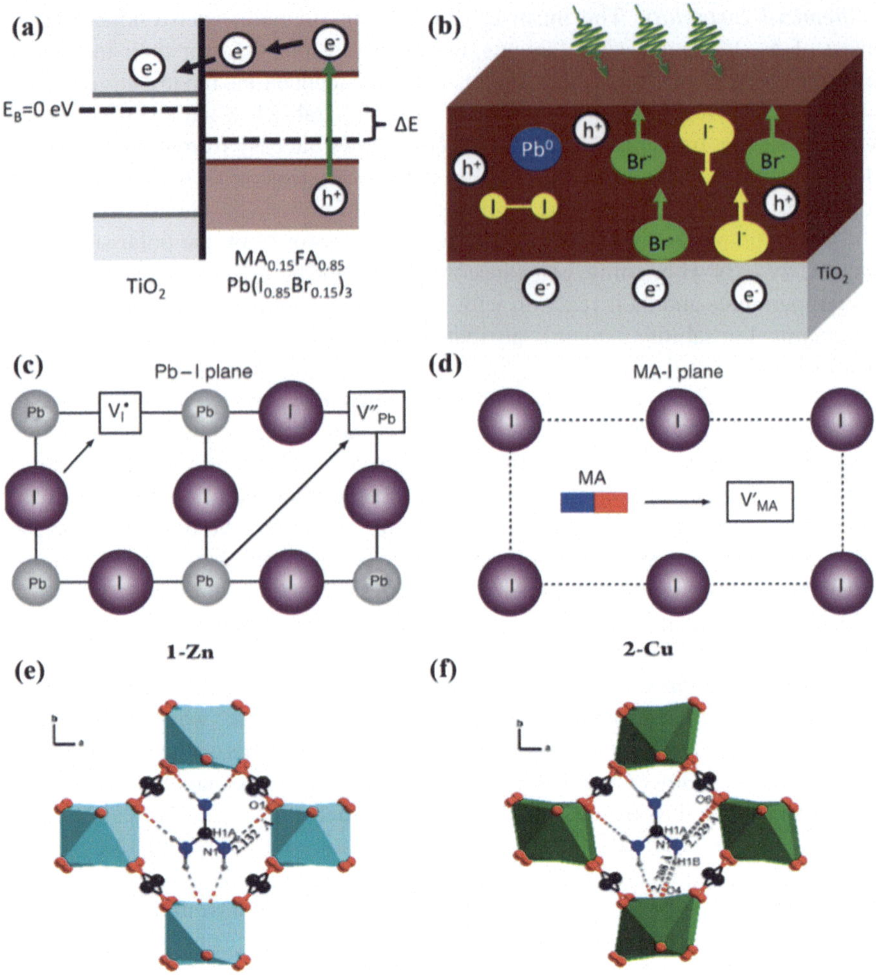

Fig. 3.4 **a** Schematic display of charge separation on illumination and consequent changes in the fermi-energy, and **b** formation of Pb^0 and I_2 due to illumination and Br-ions moving toward the surface to make it rich Br^-, Reproduced from Ref. Cappel et al. (2017), Copyright (2017) with permission of the American Chemical Society. **c, d** Schematics of ionic transport mechanism: I^- migration toward octahedron edge, Pb^{2+} migrating toward diagonal <110>, and MA^+ migrating toward A-site vacancy, Reproduced from Ref. Eames et al. (2015), Copyright (2015) with permission of Springer Nature. **e, f** Structural framework of 1-Zn and 2-Cu: where N–H···O bonds are depicted as dashed lines, while the two thicker dashed lines emphasize the elongated H···O bonds induced by the Jahn–Teller (JT) effect in 2-Cu, Reproduced from Ref. Gui et al. (2018), Copyright (2018) with permission of the American Chemical Society

most prominent electron acceptors are I_i, MA_{Pb}, V_{MA}, V_{Pb}, I_{MA}, and I_{Pb}, and electron donors are MA_i, Pb_{MA}, V_I, Pb_i, MA_I, and Pb_I. Vacancy defects having superficial levels (e.g., V_I, V_{MA}, V_{Pb}) are prevalent because of their low formation energy (Ran et al. 2018). Grain boundaries and surfaces are crucial in accumulating charged defects, capable of capturing photogenerated free carriers and acting as non-radiative recombination sites, negatively impacting the performance and stability of PSCs (Sherkar et al. 2017). However, the ionic nature of perovskites allows for the effective passivation of charged defects using extra ions like Lewis acids, Lewis bases, and other additives without altering the perovskite lattice's fundamental bonding nature. Additionally, some chemicals assist in creating superior perovskite films with bigger grains, more stable photoactive phases, and fewer grain boundaries (Fig. 3.5) (Zhang et al. 2020). Some examples of how additives can improve the operating stability of PSCs are discussed in the following paragraphs.

3.4.1.1 Defect Passivation

Previous studies have shown that incorporating additives, such as I_2, during perovskite fabrication can effectively passivate defects and enhance the optoelectronic performance of PSCs. By disproportioning I_2 and MA, Liu and his research team found that adding MA/EtOH to the perovskite ink can minimize I_2 impurities. MA can also interact with Pb^{2+} ions as a Lewis base, which limits the formation of I^- rich iodo-plumbate complexes and results in a champion PCE of 20.02% (Liu et al. 2017).

Additionally, it is found that the alkalinity of additives can have a significant impact on the kinetics of crystallization and the disproportionation reaction of I_2. While "weak" alkalis preserve the photoactive phase during crystal formation, "strong" alkalis quicken the disproportionation reaction. Formamidine acetate (FAAc), a mildly alkaline supplement containing the FA^+ cation, can act as a "residual" free agent to reduce deep-level defects in perovskites, hence improving PCEs by up to 21.01% (Chen et al. 2019b). These results indicate that incorporating suitable chemical additives during perovskite fabrication is a viable strategy to minimize defects and improve the functional durability of PSCs.

In addition to dual-passivation, Lewis acids or bases can be effectively used to passivate charged defects in PSCs. Electron-rich defects can be passivated by Lewis acids, which function as electron acceptors. Metal ions and fullerene derivatives are the two main types of Lewis acids. Through the formation of ionic bonding, positively charged metal ions can passivate negatively charged defects. Abdi-Jalebi et al. demonstrated that adding K + to a perovskite selectively removed bromide from the crystal structure, immobilizing KBr at grain boundaries and surfaces and inhibiting non-radiative decay and halide migration. Unencapsulated PSCs preserved more than 80% of their initial PCE after 300 h of maximum power point tracking (MPP) with 1-Sun irradiation in a nitrogen atmosphere (Fig. 3.6a, b). With high electron affinity, fullerene derivatives, including PCBM and C60, can serve as materials for interfacial alterations, electron transport layers, and defect polishing between perovskite layers

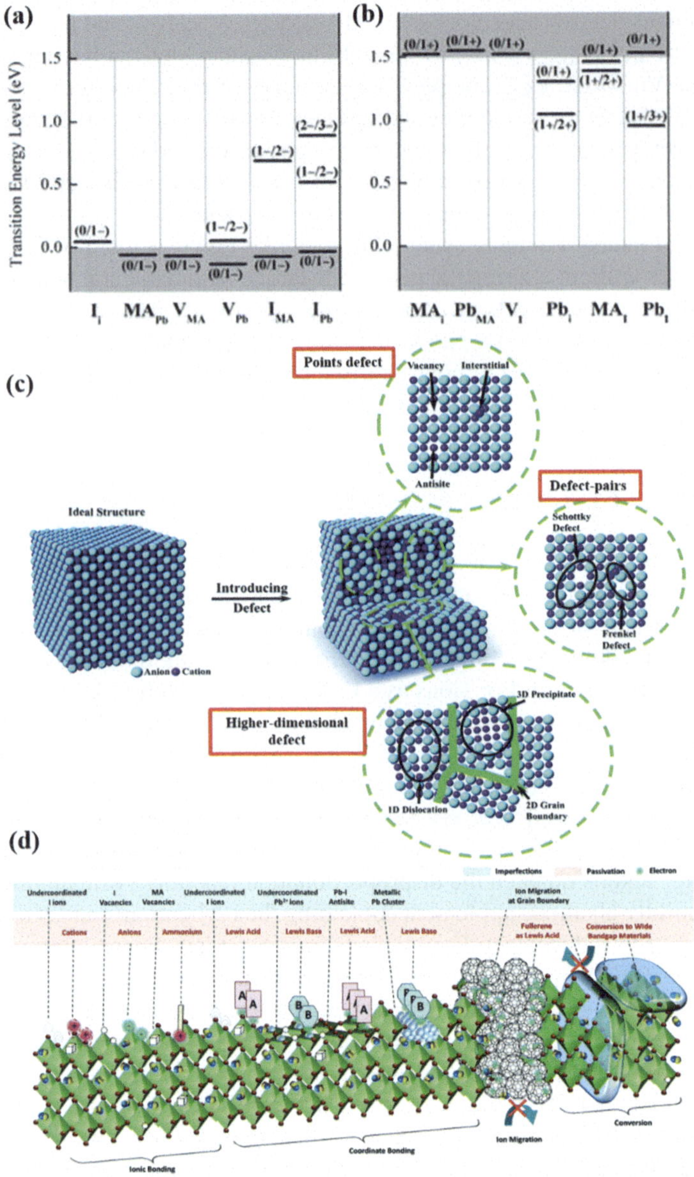

Fig. 3.5 **a**, **b** Transition energy levels of intrinsic acceptors and donors in $MAPbI_3$, respectively, Reproduced from Ref. Yin et al. (2014), Copyright (2014) with permission of AIP Publishing. **c** Schematic representation of possible defects existing in Perovskite crystal such as point defect, defect pairs, and high-dimensional defects, Reproduced from Ref. Ran et al. (2018), Copyright (2018) with permission of the Royal Society of Chemistry. **d** Typical summary of all the possible defects present in OIHP films and their passivation strategies, Reproduced from Ref. Chen et al. (2019a), Copyright (2019) with permission of the Royal Society of Chemistry

(Abdi-Jalebi et al. 2018). Xu et al. used DFT studies to show that while thermodynamically favorable bonding of PCBM with PbI_3^- inhibited the development of deep defect states, introducing PCBM close to PbI_3^- anti-site traps hybridized the ground state wave function among PCBM and the perovskite surface. The resulting PSCs had minimal hysteresis and enhanced photovoltage (Xu et al. 2015). Zhang et al. adopted α-bis-PCBM to effectively block the creation of voids/pinholes in the perovskite active layer, thereby promoting perovskite growth and film development. After 600 h of MPP tracking in an N_2 atmosphere at room temperature and with standard irradiance, PSCs built with this technology retained 96% PCE. Conversely, under the same conditions, control PSC devices lost all PCE drastically, after 250 h. (Fig. 3.6c, d). These results show that Lewis acid passivation has the potential to increase PSCs' functional stability (Zhang et al. 2017).

Lewis bases are another type of additive that can be used to passivate electron-poor defects. These bases may interact with ions that are positively charged such as Pb^{2+} and generally act as donors of electrons. Such coordination can result

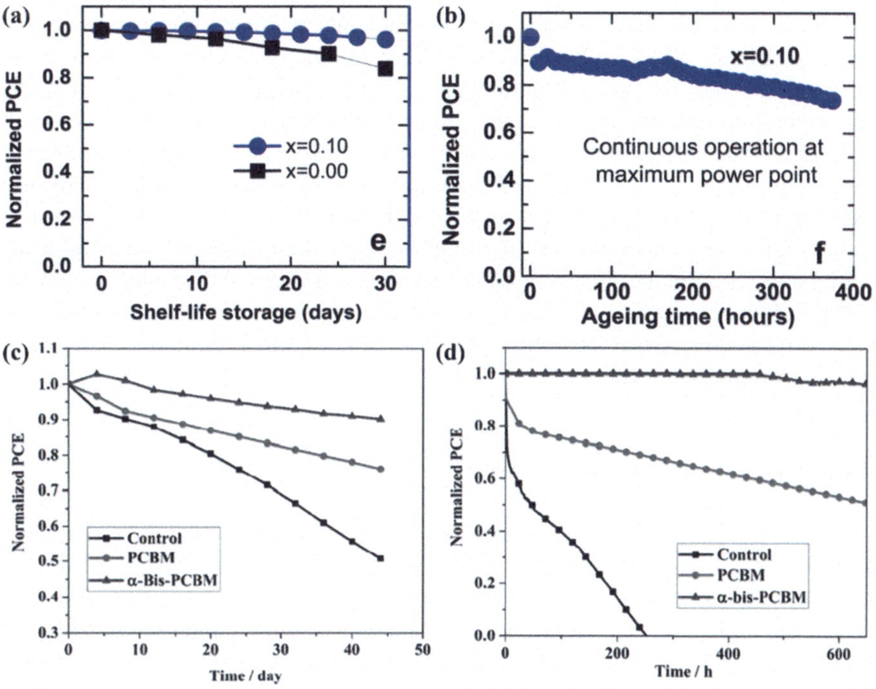

Fig. 3.6 **a, b** In a nitrogen atmosphere, unencapsulated PSCs maintained more than 80% of their initial PCE after 300 h of MPP tracking under 1-Sun illumination, Reproduced from Ref. Abdi-Jalebi et al. (2018), Copyright (2018) with permission of Springer Nature. **c, d** At ambient temperature and under 1-Sun illumination for 600 h, α-bis-PCBM aided PSC maintained 96% of their initial PCE in a nitrogen atmosphere, Reproduced from Ref. Zhang et al. (2017), Copyright (2017) with permission of WILEY-VCH Verlag GmbH & Co. KGaA, Weinheim

in acid–base compounds that passivate the electron-poor defects. There are three main types of Lewis bases: donors that are N-based, O-based, and S-based. N-based donors, such as pyridine, 4-tert-butylpyridine, D-4-tert-butylphenylalanine, 1H, and 1H-perfluorooctylamine, among others, have a nitrogen atom with a single pair of electrons. For instance, Wu et al. demonstrated in 2019 that 1,3-diaminopropane (DAP) possessing $-NH_2$ tails, which could create coordinating bonds with under-coordinated Pb^{2+} to passivate structural defects at the surface of the perovskite layer. Additionally, by exposing the hydrophobic carbon chain that links the grains and creating a barrier that keeps moisture out, it may fill A-site vacancies at the surface of perovskite grains. As a result, the encapsulated PSCs maintained 90% of their starting PCE following 500 h under 1-Sun irradiation. On the other hand, the PCE of the DAP-free device dropped to only 35% of its initial PCE after 200 h in a comparable environment. In addition, substances having a single pair of electrons bearing oxygen, as caffeine, ethyl cellulose, trimesic acid, and N-methyl-2-pyrrolidone, are known as O-based donors.

Conversely, S-based donors, such thiophene, thiourea, and 2-pyridyl thiourea, have a lone pair of electrons that bear sulfur. By bonding to under-coordinated Pb^{2+} ions, these molecules have been shown to efficiently passivate positively charged defects, producing PSCs with increased PCE and thermal stability. However, more investigation is needed to ascertain how these O or S-based Lewis, contingent on the PSCs' operating stability, influence the choice of more effective additions (Wu et al. 2019). In conclusion, the approaches for reducing defects in PSCs by incorporating additives primarily rely on creating efficient chemical reactions. Examples are the use of redox processes to eliminate certain defects, the immobilization of ions to diminish vacancy defects by chemical bonding, and the formation of acid–base complexes to efficiently coordinate the passivation of charged defects. All of these chemical methods offer advantages in ensuring long-term operational stability of PSCs. To further improve the operating stability of these devices, however, more strategies at the quantum level are required.

3.4.1.2 Morphological Engineering

Modifying the perovskite film morphology using additives efficiently enhances PSCs' optoelectronic performance and operational stability. This can involve increasing grain size, modifying crystal orientation, and minimizing film pinholes. Acetonitrile in the precursor solution can promote perovskite grain growth by facilitating solvent removal. At the same time, MACl is a widely used "film quality modification" that can increase grain size and reduce the grain boundaries in perovskite films. Tavakoli and team adopted MACl as a "crystallization aid" that increases the grain size up to 2 μm for mixed A-site perovskites. They also employed iodine solution as a post-passivator to minimize surface defects caused by the departure of chlorine atoms from the perovskite lattice upon annealing. The resulting PSCs demonstrated high operational stability at both room temperature and an elevated temperature of 65 °C and excellent retention of its initial PCE following 500 h of MPP under standard irradiation of 1-Sun in an N_2-environment. Figure 3.7a, b shows the enlarged grain size achieved by using MACl as a crystallization aid, while Fig. 3.7c demonstrates the high stability and PCE retention of the resulting PSCs (Tavakoli et al. 2019).

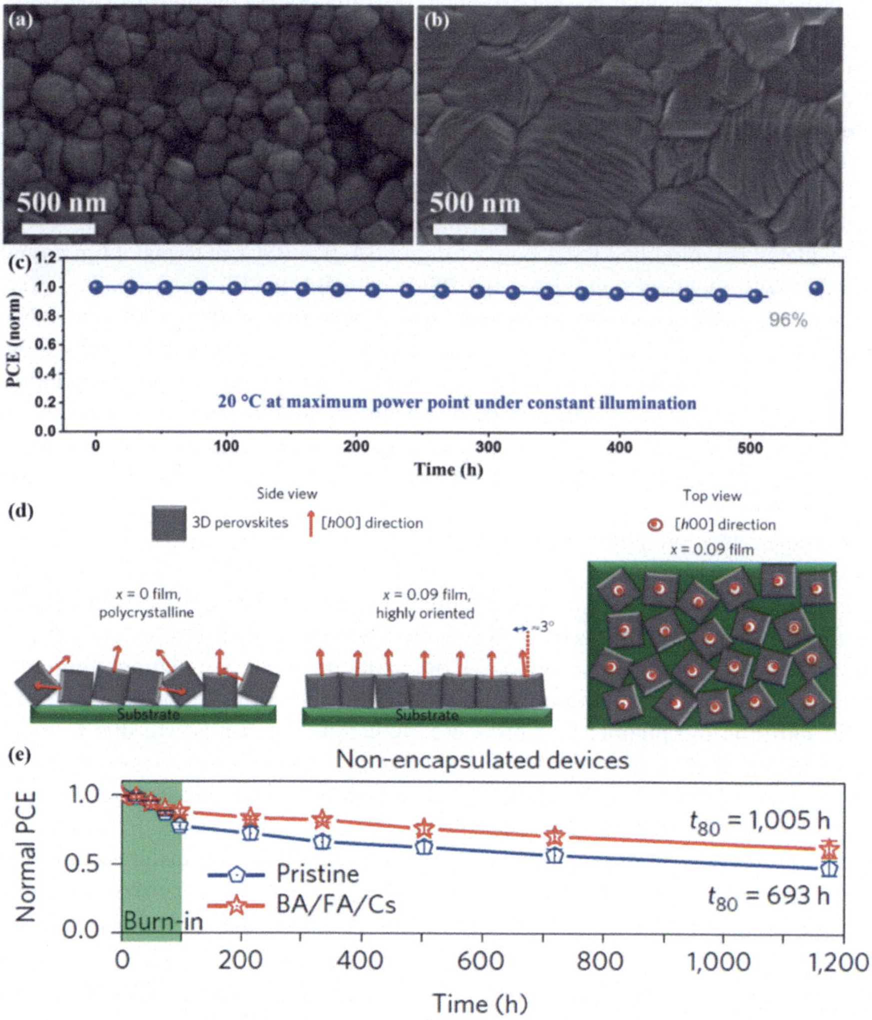

Fig. 3.7 **a**, **b** Top-view of FESEM images: Without and with MACl added triple A-cation PSC, **c** the high stability and PCE retention of the resulting PSCs, Reproduced from Ref. Tavakoli et al. (2019), Copyright (2018) with permission of WILEY-VCH Verlag GmbH & Co. KGaA, Weinheim. **d** depicts the out-of-plane alignment of the 3D perovskite grains achieved by adding BA cations, and **e** illustrates the more excellent operational stability of the BA-modified PSCs compared to the BA-free PSCs. Reproduced from Ref. Wang et al. (2017b), Copyright (2017) with permission of Springer Nature

Various approaches have been investigated to adjust the crystal orientation within perovskites. One method involves cation cascade doping at the A-site, an efficient way to control the facet orientation in perovskites. Adjusting the FA/MA ratio was also shown to prevent the thermodynamically favored crystal orientation finely. Another strategy is introducing large cations, like n-butylammonium (BA), to form 2D perovskites. A strong preference for the [100] direction for the out-of-plane alignment of the 3D perovskite grains was observed when Wang et al. added BA cations to mixed-cation lead mixed-halide 3D perovskites, "plate-like" 2D-phase crystallites formed and adhered between the grains of the 3D perovskite. The BA cations had been removed from the crystallizing 3D perovskite domains during crystallization, serving as a flux to raise the degree of crystallinity in 3D perovskites and speed up crystal growth. Improved crystal alignment has been shown to benefit PSC performance. Under 76 $mW\ cm^{-2}$ of radiation and in open-circuit conditions with 45% RH, encapsulated BA-modified PSCs outperformed BA-free PSCs in terms of functional stability. With identical conditions, the PSCs with BA modification retained 80% of their initial PCE (save for the burn-in zone) even after 1680 h, while the PSCs without BA maintained only 80% of their initial PCE for 970 h. These results show that altering the crystal orientation can significantly increase the operational stability of PSCs. Figure 3.7d depicts the out-of-plane alignment of the 3D perovskite grains achieved by adding BA cations. In contrast, Fig. 3.7e illustrates the more excellent operational stability of the BA-modified PSCs compared to the BA-free PSCs (Wang et al. 2017b).

To reduce film pinholes in perovskite films, methimazole (MMI) was used by Liu et al. to form a surface "patch" that converts residual PbI_2 at grain boundaries (GBs) into an MMI-PbI_2 complex, effectively reducing pinholes and suppressing ion migration and metal electrode diffusion. This resulted in PSCs that retained 80% of their initial efficiency after 672 h of continuous 1-Sun irradiation. Similarly, additives are widely used to improve film quality for Pb–Sn-based perovskite films, where uniform and pinhole-free films are challenging to synthesize due to rapid crystallization (Liu et al. 2018). Guanidinium thiocyanate (GuASCN) was added to $(FASnI_3)_{0.6}(MAPbI_3)_{0.4}$ by Tong et al., which significantly decreased the densities of small grains and pinholes and enhanced the compactness of GBs. In addition to improving film morphology and increasing grain size, SCN-ions also restricted the formation of excess Sn vacancies, reduced oxygen diffusion into perovskite grains to block Sn^{2+} from oxidizing further, and formed a two-dimensional structure at GBs that stopped Sn from diffusing out of grains. Consequently, after 115 h of MPP tracking under 1-Sun irradiation, encapsulated PSCs with a small bandgap of around 1.25 eV retained 90% of their initial PCE. However, under the same aging conditions, GuASCN-free PSCs only retained 80% of their initial PCE after 8.5 h. Therefore, improving film quality is critical for obtaining more stable PSCs during operation (Tong et al. 2019).

As discussed earlier, the inadequate morphology of perovskite films can generate defects that act as non-radiative recombination sites for free carriers, causing a significant decline in the optoelectronic performance of PSCs. In addition, ion migration inside perovskites may have a detrimental effect on PSC long-term stability by causing lattice deformation and a transition into a photoinactive phase. The use of functionalized additives, however, can help with a few of these problems. This can be achieved by (i) engineering appropriate chemical processes inside perovskites to passivate defects or adding extra chemical bonds, (ii) modifying the morphology by regulating the kinetics of perovskite nucleation and growth, and (iii) enhancing the energy barrier of phase transition and strengthening the cohesiveness of perovskite structures to stabilize the photoactive phase of perovskites, as presented in Table 3.2. These chemical methods have demonstrated efficacy in enhancing the stability of PSCs. However, additional developments from a physical or atomic standpoint are needed to reach even higher levels of stability in real-world operational scenarios.

3.4.2 Cation/Anion Engineering

Perovskites are distinguished by their versatile composition in the A, B, and X-sites. As long as the ions' sizes meet the Goldschmidt tolerance factor calculation requirements, these sites can accommodate single or mixed ions. Moreover, the application of large monovalent cations on the A-site, such as n-butylamine (BA) or phenyl-ethyl-ammonium (PEA), may produce 2D or quasi-2D perovskites, which are the subject of extensive study. As perovskites have different components, each having distinct optoelectronic properties, component engineering has a vast potential to enhance the PSC's operational stability. Even though there is no systematic comparison to determine which component is a better option for operation, the summary that follows, depending on various types of perovskite components, can be helpful in selecting more effective, stable perovskites.

3.4.2.1 A-Site Cation Modification

Despite the A-cation having no direct impact on the electronic properties of perovskite, simulations have demonstrated that the size of the A-cation can affect the degree of distortion, which in turn influences the electronic characteristics of the material. An asymmetrical three-dimensional structure can be achieved by introducing a small monovalent cation like cesium, rubidium, methylammonium, or formamidinium to occupy the A-site vacancy. These results emphasize how crucial it is to consider the size and type of cation used in perovskite materials to optimize their electronic properties.

Methylammonium ion ($CH_3NH_3^+$ or MA^+) is a commonly used cation in hybrid perovskites and has enabled perovskite devices to achieve PCEs of 15% and above for $MAPbI_3$ devices. However, because of the smaller size of the MA cation, $MAPbI_3$ forms a tetragonal symmetry, resulting in a bandgap of 1.51–1.55 eV, which is more significant in comparison to that of optimal Shockley-Queisser limit for a mono-junction device. To improve light absorption across the spectrum, researchers have investigated the use of larger cations, like ethyl-ammonium (EA), formamidinium

Table 3.2 Typical summary of lifetime testing under different operational conditions

Device structure	Enhancer	Test parameters under MPP tracking	PCE retention	References
FTO/c-TiO_2/m-TiO_2/ $Cs_{0.06}FA_{0.79}MA_{0.15}Pb(I_{0.85}Br_{0.15})_3$/ spiro-MeOTAD/Au	K^+	At room temperature, 1-Sun illumination, N_2 environment	300 h @80%	Abdi-Jalebi et al. (2018)
ITO/PTAA/$MAPbI_3$/C_{60}/BCP/Cu	1,3-Diaminopropane	Real application scenario, 1-Sun illumination	500 h @90%	Wu et al. (2019)
ITO/SnO_2/ $Cs_{0.05}FA_{0.54}MA_{0.41}Pb(I_{0.98}Br_{0.02})_3$/ spiro-MeOTAD/Au	Fluoride	At room temperature, 1-Sun illumination, N_2 environment	1000 h @90%	Li et al. (2019)
FTO/TiO_2/SnO_2/(FA,MA,Cs)Pb(I, Br)$_3$/ spiro-MeOTAD/Au	MACl	At 20 °C, 1-Sun illumination, N_2 environment	500 h @96%	Tavakoli et al. (2019)
FTO/SnO_2/C_{60}/ $BA_{0.09}(FA_{0.83}Cs_{0.17})_{0.91}Pb(I_{0.6}Br_{0.4})_3$/ spiro-MeOTAD/Au	BA^+	76 mW cm^{-2}, 1-Sun illumination, 45% RH	1680 h @80%	Wang et al. (2017b)
FTO/c-TiO_2/m-TiO_2/$FAPbI_3$/ spiro-MeOTAD/Au	$MDACl_2$	At room temperature, 1-Sun illumination, Ambient conditions	600 h @90%	Min et al. (2019)

(FA), and cesium (Cs), where the cation radius follows the order $r_{Cs} < r_{MA} < r_{FA} < r_{EA}$. However, replacing MA with EA resulted in a two-dimensional orthorhombic crystalline morphology with a higher bandgap of 2.2 eV for $EAPbI_3$ and a comparatively low efficiency of 2.4% under 1-Sun illumination (Koh et al. 2014). Hence, it is necessary to investigate organic cations within the radius ranges from MA to EA to optimize hybrid perovskites' electronic properties. The formamidinium cation ($HC(NH_2)_2^+$ or FA), with its larger size than the methylammonium ion, is expected to provide a better symmetry in hybrid perovskites and has therefore been thoroughly explored. $FAPbI_3$ crystals exhibit a desirable bandgap value of approximately 1.43–1.48 eV, with absorption beginning around 840 nm, close to the ideal value of 1.4 eV. However, it has often been noted that $FAPbI_3$ can form a yellowish hexagonal non-perovskite polymorph, which is anticipated to reduce photovoltaic performance because of poor electron injection resulting from unfavorable band alignment with TiO_2. The total removal of this non-perovskite polymorph state could give rise to higher efficiency for devices based on FA-based devices compared to those found on MA. Scaife et al. were the pioneers in investigating cesium perovskites, and they discovered that $CsSnI_3$ emits strong photoluminescence in the near-infrared region at room temperature. Chung et al. identified four distinct structural phases of $CsSnI_3$ perovskites that exist independently under ambient conditions. The relatively small size of Cs-atoms compared to MA atoms causes octahedral tilting, which leads to a broader bandgap (1.73 eV) in $CsPbI_3$ than in $MAPbI_3$ (1.57 eV), surpassing the optimal bandgap of 1.4 eV more significantly. Mixed A-cation perovskites leverage the favorable properties of their constituent parts. Pellet et al. demonstrated the tunability of the bandgap of $MA_xFA_{1-x}PbI_3$ perovskites by adjusting the ratio of MA to FA cations, attaining a remarkable efficiency of 14.9%. This enhanced efficiency is attributed to the improved absorption of the red region of the spectrum. Mei et al. developed PSCs based on 5-amino valeric acid $(5\text{-}AVA)_xMA_{1-x}PbI_3$, which exhibited reduced defects and superior compatibility with the adjacent titania film. This allowed for a longer exciton lifespan and increased charge separation compared to pure $MAPbI_3$. These devices surpassed pure MA perovskites in terms of efficiency and stability, achieving 12.8% under normal environmental conditions and light.

The initial focus of A-site component engineering in perovskite PV technology was to enhance the optoelectronic performance of PSCs, especially PCE, by switching from $MAPbI_3$ to $FAPbI_3$ due to its more suitable bandgap. However, the α-phase of $FAPbI_3$ is thermodynamically unstable at room temperature and prone to transform into the yellow, photoinactive δ-phase, which severely hampers the design of efficient and stable $FAPbI_3$-based PSCs. In order to overcome this problem, scientists have looked to FAMA systems, which show a more stable black phase and are hence a good option for high-efficiency and stable PSCs. Nonetheless, a methodical examination of the functional reliability of FAMA alloys and perovskites that solely contain MA under identical aging circumstances has not yet been carried out (Jiang et al. 2018).

In 2016, Saliba et al. systematically studied the impact of incorporating cesium (Cs) into the frequently employed FAMA perovskite system $MA_{0.17}FA_{0.83}Pb(I_{0.83}Br_{0.17})$. For FA-based perovskites, Cs acts as a "phase stabilizer" for the black phase due to its reduced ionic radius. The addition of only 5% Cs was found to efficiently prevent the development of the yellow phase of perovskites, which is attributed to Cs inducing perovskite seeds to assist film formation. XRD results indicated a distinct perovskite peak in the untreated CsFAMA perovskite film. Unencapsulated PSCs containing 5% Cs demonstrated a notable improvement in functional stability through phase stabilization and grain alteration, maintaining 90% of the initial PCE after 250 h of MPP tracking under continuous 1-Sun illumination in an N_2 environment at room temperature (Saliba et al. 2016). In contrast, during the first 100 h, PSCs based on FAMA dropped more than 50% of their PCE. In contrast, $FAPbI_3$ only kept about 20% of the original PCE after 4 h of the same conditions. Jiang et al. found that an unencapsulated $Cs_{0.07}FA_{0.93}PbI_3$-based PSC retained 56% of its initial PCE after 17 h of MPP tracking under 1-Sun illumination with 60% RH. In addition to the positive effect of Cs in stabilizing the black phase, the study also showed that the energy levels of perovskites with Cs being incorporated were precisely matched with TiO_2 and spiro-OmeTAD, allowing carrier movement. These results indicate that Cs ions positively impact stabilizing the black phase, changing grain morphology, and regulating energy levels of perovskites, thereby promoting PSCs to achieve improved functional stability.

Saliba et al. investigated the impact of rubidium (Rb) addition in the formamidinium-based cesium lead halide perovskite system and the formamidinium-based perovskite system. Rb was found to preserve the black phase of perovskites based on FA by removing excess Pb and contaminants from the yellow phase, which resulted in a uniform crystallization upon annealing, despite the fact that Rb has a smaller ionic size than Cs and cannot form the black perovskite phase $RbPbI_3$. Additionally, incorporating 5% Rb in RbCsFAMA led to larger grain sizes of perovskites than CsFAMA. PSCs using PTAA and RbCsFAMA as the HTL preserved 95% of their original PCE after 500 h of MPP tracking at 85 °C in a N_2-environment with constant 1-Sun irradiation (Kim et al. 2017). Despite the demonstrated benefits of incorporating Rb and Cs into FA-based PSCs, there is a lack of research investigating the impact of Rb alone on operational stability. Moreover, there is still debate on whether Rb occupies the A-site in perovskites and the underlying mechanism of the improved performance following Rb addition.

3.4.2.2 B-Site Cation Modification

Hybrid perovskite materials typically incorporate group IVA metals in a +2-oxidation state (Pb^{2+}, Sn^{2+}, Ge^{2+}) in the M metal cation position. Lead (Pb) has been the most widely utilized, outperforming tin (Sn) in performance and stability. Germanium (Ge) has been less studied due to its unstable behavior in the +2-oxidation state. Along the periodic table, the stability of the +2-oxidation state goes down from Pb to Ge due to a reduced inert electron pair effect, but the electronegativity difference or ionic character goes up, causing a bandgap reduction. Although $MASnX_3$ theoretically displays a more suitable bandgap of 1.2–1.4 eV compared to that of $MAPbX_3$ (1.6–1.8 eV), it rapidly oxidizes in air to Sn^{4+}, producing a volatile SnI_4 compound (Lee et al. 2017). Hao et al. developed hybrid perovskites that are lead-free, like $MASnI_3$ and $MASnI_{3-x}Br_x$, achieving an efficiency of 5.73%, which was later improved to 6% by Noel et al. under one sunlight intensity. Stoumpos et al. discovered that although $MASnI_3$ and $MAPbI_3$ adopt a tetragonal arrangement (P4mm) under normal conditions, $MASnI_3$ assumes a better-conforming Γ-state while $MAPbI_3$ takes a β-state. While $MASnX_3$ perovskites possess the potential to produce a greater short-circuit current density (J_{SC}) in comparison to $MAPbX_3$, their instability has impeded further development.

The bandgap of $MASn_{1-x}Pb_xI_3$ perovskite devices was significantly adjusted by modifying the ratio of Pb to Sn. By enhancing the Sn content up to $x = 0.5$, a quasi-cubic tetragonal P4 mm structure identical to unadulterated $MASnI_3$ is achieved. The arrangement transforms to a tetragonal I4 cm structure as the amount of Sn increases, leading to enhanced absorption of photons and a decrease in bandgap. Dual-metal hybrid perovskite devices have significantly improved, as discussed above.

Sn can efficiently substitute Pb in perovskite materials and lower the bandgap, ranging from 1.3 eV for $MASnI_3$ and $CsSnI_3$ to 1.41 eV for $FASnI_3$. This is beneficial for the absorbing of solar radiation and makes Sn-based perovskites attractive for use as rear cells in tandem photovoltaic devices, as their bandgap can be adjusted between 1.16 eV ($MASn_{0.6}Pb_{0.4}I_3$) and 1.41 eV ($FASnI_3$). However, the stability of Sn-containing PSCs must be considered. Unencapsulated PSCs with $FASnI_3$ lost their entire PCE within minutes due to the air sensitivity of Sn-based films and the oxidation of Sn^{2+} to Sn^{4+}. However, PSCs based on $FA_{0.75}Cs_{0.25}Pb_{0.5}Sn_{0.5}I_3$ and $FAPb_{0.5}Sn_{0.5}I_3$ retained more than 90% of their initial PCE after 50 min, suggesting improved operational stability. When exposed to ambient air with an RH of 50 $\pm$ 5% and 50 min of MPP tracking, Sn-containing PSCs retained over 90% of their initial PCE, indicating that their operational stability was superior to that of $FAPbI_3$-based devices. This improved stability may be due to the partial inhibition of Sn^{2+} oxidation by Pb–Sn alloys; it emphasizes the viability of Pb–Sn-based PSCs for practical application and raises the prospect of a chemical reaction between Pb and Sn ($Pb + Sn^{4+} = Pb^{2+} + Sn^{2+}$). The exact mechanism underlying this phenomenon is currently being investigated. This result also indicates that the black-to-yellow phase transition described in other publications for the $FAPbI_3$-based device might be associated with decreased stability.

Notably, the oxidation of Sn^{2+} remains a significant challenge in advancing Pb–Sn-based PSCs. However, Gu et al. made an essential discovery by demonstrating that incorporating Sn powder into a precursor solution containing Sn^{2+} ions may efficiently remove Sn^{4+} ions via a comproportionating reaction of Sn and Sn^{4+} to produce $2Sn^{2+}$. Pb–Sn PSCs currently exhibit inferior operational stability compared to cutting-edge Pb-based devices because of the high sensitivity of Sn^{2+} to oxidation and the poor film quality of perovskite. However, Pb–Sn alloys have a unique small bandgap that allows them to help create stable perovskite-perovskite tandem photovoltaics that are operationally stable (Gu et al. 2018).

In addition, various elements with stable oxidation states of -2 can be used as lead substitutes in the perovskite structure. These include group-14 elements such as Ge^{2+} and Sn^{2+}, alkaline-earth metals such as Be^{2+}, Mg^{2+}, Ca^{2+}, Sr^{2+}, and Ba^{2+}, transition metals like V^{2+}, Mn^{2+}, Fe^{2+}, Co^{2+}, Ni^{2+}, Pd^{2+}, Pt^{2+}, Cu^{2+}, Zn^{2+}, Cd^{2+}, Hg^{2+}, lanthanides such as Eu^{2+}, Tm^{2+}, and Yb^{2+}, and p-block elements such as Ga^{2+} and In^{2+}. However, due to their high bandgaps, toxicity, radioactivity, or instability in the −2-oxidation state, several of these elements cannot produce perovskites and cannot be suitable for photovoltaic applications. After considering these factors and using computational screening, the most promising substitutes for lead in cesium and methylammonium metal halide perovskites are Sn^{2+}, Ge^{2+}, Mg^{2+}, V^{2+}, Mn^{2+}, Ni^{2+}, Zn^{2+}, and Co^{2+} (Kour et al. 2019).

However, another approach to creating lead-free perovskite materials is through heterovalent substitution. This technique uses a cation of a different valence state, such as a monovalent, trivalent, or tetravalent cation, to replace the divalent lead cation. However, a simple substitution with heterovalent cations is impossible due to the different valence states. There are two methods for replacing heterovalent atoms. The first is the mixed-valence method, which involves using an equal number of monovalent and trivalent cations to achieve an average overall valence state of −2, similar to Pb^{2+}. Thallium and gold halide perovskites are examples of perovskites made using the mixed-valence approach. In the second technique, trivalent cations like Sb^{3+} and Bi^{3+} replace the divalent Pb^{2+}. However, this significantly transitions the perovskite structure from an ABX_3-type to an $A_3B_2X_9$-type perovskite for maintaining charge neutrality (Nie et al. 2020).

The heterovalent substitution approach has the potential to make significant advancements in the advancement of new lead-free perovskite semiconductors because it allows for the use of cations with valence states other than divalent. This opens up the possibility for further experimentation and exploration of different cations to improve the properties of lead-free perovskite semiconductors. We will see all these substitutions for making PSCs stable and enhancing their efficiency in the next chapter, where we broadly discuss the variety of possible substitutions of Pb to make this commercially viable.

3.4.2.3 X-Site Anion Modification

The most efficient method for investigating the functional characteristics of hybrid perovskite photovoltaics is by varying the halide ion. It has been observed that moving down the VIIA group causes the atomic size to increase, which causes the absorption spectra to shift toward longer wavelengths and the energy to decrease.

Hybrid perovskite devices based on iodide ions have achieved the highest efficiencies, with values exceeding 20%. Iodide has an electronegativity value (2.66 EN) that is closer to Pb (2.33 EN), resulting in a bond character that is neither purely ionic nor purely covalent but rather a mix of both, which contributes to the stability of the perovskite structure. However, the stability of iodide-based perovskites toward humidity is a serious problem, which has prompted further research to explore alternatives to iodide and mixed-halide perovskites. The addition of chloride (Cl^-) ions to perovskite materials has been shown to improve carrier lifetimes and diffusion lengths, leading to increased efficiency in planar perovskite devices. While $MAPbCl_3$ has a cubic structure at room temperature, blending it with other ions, such as $MAPbI_{3-x}Cl_3$, can result in a more aligned crystalline structure. XRD analysis of $MAPbI_{3-x}Cl_3$ has shown that the composition of chlorine-based perovskites is similar to that of iodine-based perovskites, with a maximal 3–4% chlorine-to-iodine ratio. Some studies have suggested that chlorine sublimes and transforms into pure chlorine gas, while others have shown that it remains unconverted and acts as nucleation vacancies, which improves the perovskite layer's surface masking. It is proposed that the development of complex ion agglomerates during crystallization, because of the limited solubility of precursor materials comprising chlorine, enables the removal of excess organic components and facilitates the development of perovskite crystals and grains. Therefore, incorporating chlorine can control the morphological evolution of perovskite films and contribute to tuning their characteristics and solar cell efficiency (Ansari et al. 2018).

Fluoride (F) is a promising option for photovoltaic applications due to its strong electron-withdrawing nature and the potential to form stronger hydrogen bonds (N–H–F) with nitrogen atoms than various halides. However, the use of fluoride is limited due to its smaller size than iodide, which causes strain in the lattice and reduces its light-absorbing and conducting abilities. In recent research, the molecular ion BF_4^- was substituted with X to create the hybrid perovskite $MAPbI_{3-x}(BF_4)_x$. Under AM1.5 light intensity, the substituted perovskite outperformed $MAPbI_3$ regarding electrical conductivity and photon response. The strong electronegativity of the F constituent in the BF^{4-} ion forms a firm hydrogen bond with the MA^+ ion, resulting in a lower degree of ionization of MA^+.

Bromide (Br) is effective in adjusting the bandgap of perovskites. Adding Br into iodide-based perovskites induces a structural distortion that widens the bandgap, as proposed by Moscon et al. $MAPbBr_3$ adopts a cubic structure with large crystals at room temperature. The efficiency and stability of $MAPbI_xBr_{3-x}$ perovskite can be enhanced by tailoring the bandgap via chemical combination. In mixed-halide $FAPbI_xBr_{3-x}$ perovskite, Eperon et al. showed that altering the I: Br component ratios results in an adjustable bandgap ranging from 1.48 to 2.23 eV with a variety of film colors. They noticed that increasing the iodide composition causes a red shift in the bandgap and induces a transition from a cubic Br-rich state to an I-rich tetragonal state. The maximum efficiency obtained was 14.2%, with a J_{SC} exceeding 23 mA cm^{-2}. Moreover, increasing the Br ratio improved the open circuit voltage (V_{OC}) and reduced the J_{SC} (Jacobsson et al. 2016).

Early studies showed that incorporating Br into MA-based perovskites resulted in tunable bandgaps ranging from 1.5 to 2.3 eV. Additionally, due to $MAPbI_3$'s compact and stable structure, introducing a small quantity of Br improved the opto-electronic characteristics and moisture stability of related PSCs. The substitution of small bromine (Br) for large iodine (I) resulted in stronger hydrogen bonds between organic cations, halogen and lead, as well as a decrease in the lattice constant. To create stable, high-performing PSCs, a little amount of Br-ions, such as $I_{0.85}Br_{0.15}$, have been utilized extensively. However, some studies reported phase separation in the hybrid halide system ($I_{1-x}Br_x$) under irradiation that can negatively impact the operation of PSCs. Thus, a thorough analysis of the functional stability of PSCs based on several Br components is necessary in order to build more stable PSCs (Yang et al. 2018b).

According to Saidaminov et al., incorporating Cl into perovskites resulted in a reduction of the density of vacancy defects by relaxing lattice strain, in addition to a bandgap tuning effect. The BX_6 octahedra is tilted and cage distortions result from the size difference between the lead halide cage and the A-site cation, which is the source of this strain. As a result, X-site vacancies can arise, a significant cause of degradation in PSCs. However, adding small Cl ions can release the lattice strain without creating vacancy defects. This effectively prevents the development of vacancy defects, causing slower degradation rates of PSCs under illumination. Cl-containing unencapsulated PSCs degraded PCE at a slower rate (lost 0.25% per minute) during 1-Sun irradiation at roughly 50% relative humidity than Cl-free PSCs, which produced 1.5% PCE per minute. It should be highlighted that while Cl can be useful in generating consistent and stable perovskite films, more investigation is required to confirm its beneficial effects on PSCs' operating stability (Zheng et al. 2023).

In conclusion, component engineering can modify tolerance factors and residue strain to enhance the structure and stability of perovskites. It can also regulate nucleation and precursor purification to enhance the quality of perovskite films. These benefits have helped achieve greater PCE and improved the operational stability of PSC devices. For more stability enhancement, finding an appropriate perovskite component is essential. $FAPbI_3$ and $CsPbI_3$ are currently in demand because of their high efficiency and thermal stability. Due to their single-phase features, these components have a higher potential for high operational stability because they are not impacted by phase segregation during illumination. The subsequent sections will examine further effective methods for augmenting the functional stability of PSCs through the utilization of these constituents.

3.4.3 Interfacial Passivation

The interface between different materials dramatically influences the efficiency of photovoltaic devices. To achieve high efficiency, it is essential to control the behavior of carriers at these interfaces carefully. Additionally, preventing interface degradation is necessary to achieve the long-term stability of solar cells. Therefore, significant research efforts have been devoted to manipulating carrier dynamics at interfaces and designing interfaces that enhance device stability. During the solution processing of perovskite films, the transition from liquid to solid results in the formation and growth of crystals (Aharon et al. 2015). Due to the self-assembly properties of perovskites, crystal formation usually starts from the substrate and stops at the perovskite film surface. Because of this, the surface of the perovskite film contains many dangling bonds that serve as defect sites, particularly vacancy defects. Perovskite degradation typically originates from these defect sites because they are highly reactive toward water, oxygen, and other ecological factors. Moreover, during the movement of free carriers, an energy level mismatch and non-ohmic contact at the heterojunction between charge transport layers and the perovskite surface may result in energy loss. These factors collectively suggest that the perovskite surface is susceptible to degradation and can be a significant factor in the operational instability of PSCs. Therefore, to improve the stability of PSCs, it is essential to develop efficient surface passivation techniques (Wang et al. 2017a). Organic, inorganic, and low-dimensional materials are examples of surface passivation techniques that have been effectively applied to modify perovskite surfaces and enhance the functional stability of PSC devices. These several kinds of surface passivation will be the subject of the discussion that follows.

3.4.3.1 Organic–Inorganic Surface Passivation

As mentioned, charged point defects are the most prevalent defect sites in perovskites, and Lewis acids or bases can passivate them. By using molecules with charged functional groups through surface passivation, this method can also remove surface imperfections. Theophylline, caffeine, and theobromine have been shown by Wang et al. to be potential surface passivation agents for Pb_I anti-site defects. After a thorough analysis, it was shown that theophylline had the strongest Lewis base–acid interaction of the three compounds, with the C=O group and the anti-site Pb_I producing the most substantial passivation effects. Theophylline's ability to bind onto the Pb_I defect was further boosted by the hydrogen bond contact between I of the Pb–I octahedron and the nearby N–H on the imidazole ring. On the other hand, adding a methyl group to the N on the imidazole ring of theophylline (yielding caffeine) eliminated the hydrogen bond between N–H and I, weakening the interaction between the C=O groups and Pb_I defects. Additionally, when the N–H group was next to the C=O group on the same six-membered ring, the spatially effective interaction between the N–H and I was hindered (creating theobromine) because C=O had previously attached to anti-site Pb_I. Because of steric effects, this led to less effective interactions between the C=O group and Pb_I defects. The strong Lewis base–acid interaction between the C=O group and the anti-site Pb_I as well as hydrogen bonding have been found to be the mechanisms by which theophylline efficiently passivates

Pb_I anti-site defects on perovskite surfaces. Under 1-Sun irradiation and open-circuit conditions, PSCs treated with theophylline demonstrated the highest PCE of 23.48%. After 500 h with a RH of 30–40% and a temperature of about 40 °C, they preserved more than 95% of their initial PCE (Wang et al. 2019a).

In contrast, untreated PSCs experienced a significant decrease in PCE under identical aging conditions. These results suggest that both the formation of contacts between functional groups and defects, as well as the reinforcement of those bonds to produce strong and stable bonding, are necessary for successful techniques to passivate surface defects and enhance the functional stability of PSCs. It is essential to find alternative passivation methods that can target coexisting positively and negatively charged defects that cause perovskite films to deteriorate as well as passivate certain electrical defects present on the surface of perovskite films.

In addition to surface defect passivation, enhancing the stability of PSCs under actual operating conditions also requires protecting the perovskite surface from environmental factors such as water. Water molecules in the perovskite layer can be resisted by hydrophobic chemicals, although hygroscopic molecules can absorb water before it reaches the layer. The hydrophobic group of an amino group molecule (NH_2-POSS) has been demonstrated in earlier research to decrease the charge recombination density of the perovskite film and enhance the stability of PSCs under humid conditions. In 2020, a new molecule called 4-tert-butyl-benzylammonium iodide (*t*-BBAI) was developed by substituting the hydrogen of benzylammonium iodide with a hydrophobic 4-tert-butyl group. The *t*-BBAI molecule serves two purposes: hydrophobicity and defect passivation. This is because of its chemical affinity to phenyl-ethyl-ammonium iodide, which has been shown to be useful in passivating defects on perovskite films. The efficacy of utilizing *t*-BBAI molecules to enhance PSC stability was demonstrated by the unencapsulated PSC, passivated by a *t*-BBAI interlayer, which maintained 95% of PCE after 500 h under 1-Sun irradiation in a N_2-atmosphere. On the other hand, over the same duration of time and under similar conditions, the PSCs without t-BBAI interlayer only retained 70% of PCE (Fig. 3.8) (Zhu et al. 2020).

The outcomes of the studies mentioned above suggest a straightforward strategy to enhance the functional durability of PSCs in humid environment. This approach involves amphiphilic molecules that can be added to the perovskite surface. These compounds have groups that resist water molecules on one side and groups that passivate surface imperfections on the other, such as donor or acceptor groups. PMMA is an example of a polymer that exhibits similar functionality and has been demonstrated to enhance the efficiency and operational stability of PSCs successfully. Hygroscopic polymers are another category of molecules that can be used to enhance the stability of PSCs. As seen in Fig. 3.9a, Kim et al. added a thin layer of polyethylene oxide (PEO) between the charge transfer layers and the perovskite layer in PSCs. A coherent connection between PEO and the perovskite layer was formed when the lone electron pairs on PEO's, C–O shared with Pb^{2+} unoccupied 6*p* orbital. Under 1-Sun irradiation in air with 50% relative humidity, this bond passivated the positively charged

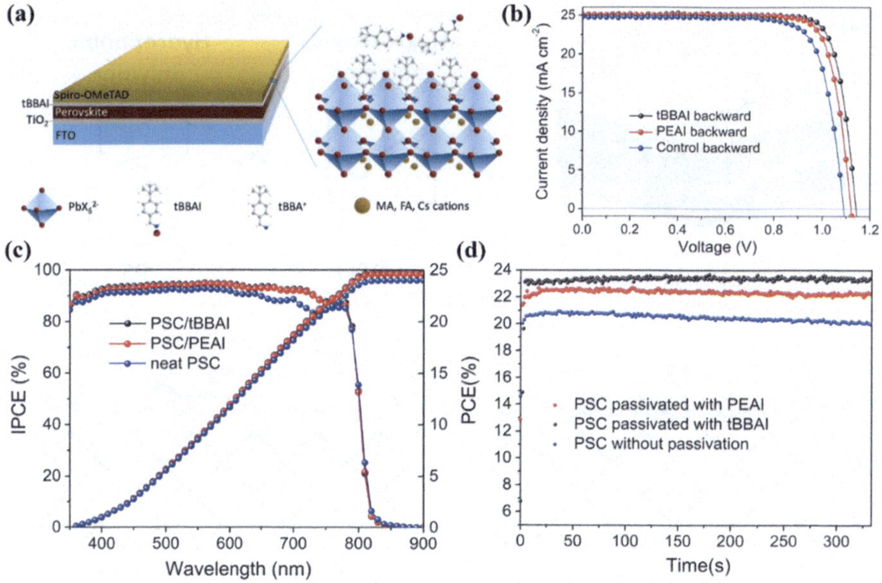

Fig. 3.8 **a** Schematic of *t*-BBAI passivated PSC, **b**, **c**, reverse *J*–*V* and IPCE analysis of *t*-BBAI and PEAI passivated PSCs, and **d** PCE stability (MPP Tracking) of passivated PSCs for first 300 h under ambient air conditions. Reproduced from Ref. Zhu et al. (2020), Copyright (2020) with permission of WILEY-VCH Verlag GmbH & Co. KGaA, Weinheim

iodine vacancy defects, leading to PEO-based PSCs that showed nearly no deterioration over 14 h of MPP tracking. Some surface treatments can also enhance the phase stability of perovskites in addition to passivating surface defects (Fig. 3.9b, c) (Kim et al. 2018b).

Beyond surface defect passivation, specific surface treatments can enhance the phase stability of perovskites. Wang et al. reported an emerging interface engineering technique utilizing choline iodine (CHI) in 2019. Apart from the previously published passivation of perovskite surface trap states, the method also created CHI shells that permeated the bulk of $CsPbI_3$ thin films. The γ-phase of the perovskite grains was hindered from changing to the photoinactive δ-phase by these shells, which also improved the quality of the perovskite film. Consequently, after 500 h of MPP tracking under 1-Sun irradiation in a N_2 glovebox, the β-$CsPbI_3$-based PSCs (unencapsulated) showed a PCE larger than 18% and retained 92% of their original PCE (Wang et al. 2019c).

While organic passivation has shown promise in enhancing PSC stability and effectiveness, their effectiveness is restricted by the feeble secondary bonding that exists between these passivation molecules and the perovskite surface. These weak bonds are insufficient in protecting the absorber from degradation caused by different aging conditions, mainly when operated in environments with moisture and oxygen. However, in silicon photovoltaic technology, materials such as silicon oxide, silicon

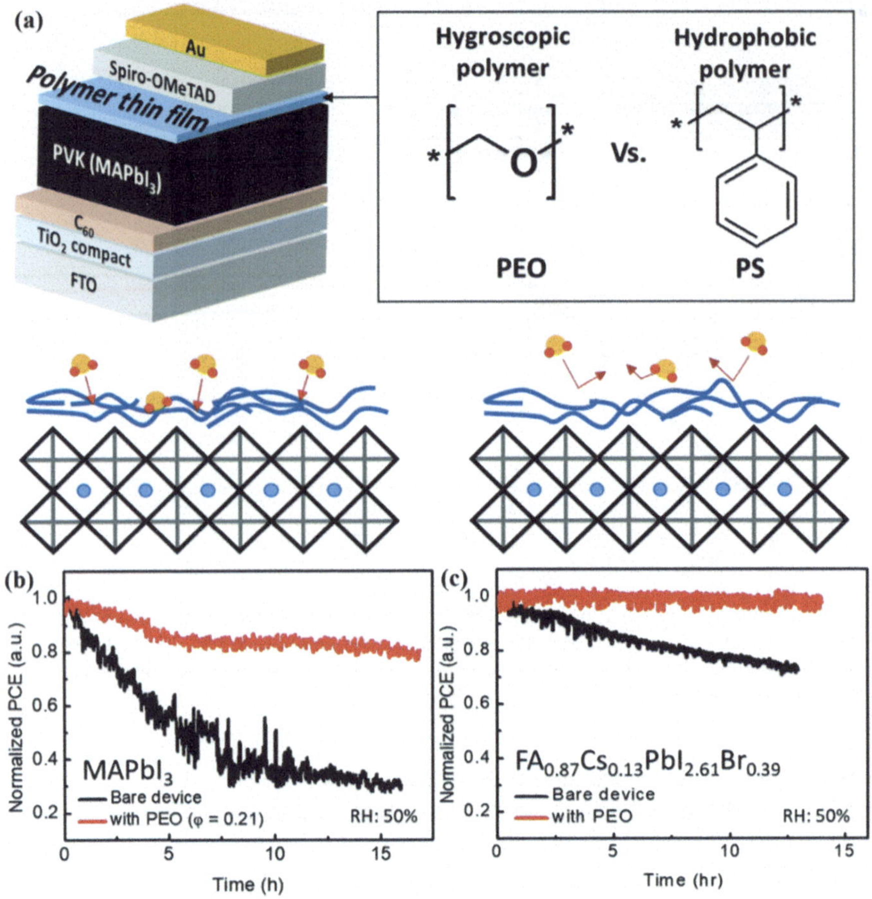

Fig. 3.9 **a** Schematic illustration demonstrates the assembly of the polymer thin film on the perovskite structure. The perovskite planar heterojunction solar cells consist of polymer thin films, specifically PEO and PS, incorporated within their device structure. **b**, **c** Operational stability tests were conducted to evaluate the PCE evolution of $MAPbI3$ and $FA_{0.87}Cs_{0.13}Pb(I_{0.87}Br_{0.13})_3$ devices in the absence and presence of PEO, respectively. Reproduced from Ref. Kim et al. (2018b), Copyright (2018) with permission of the Royal Society of Chemistry

nitride, or aluminum oxide are employed to passivate surface imperfections because they have the ability to form strong covalent connections with silicon. These inorganic coatings are appropriate for shielding the underlying silicon from deterioration without introducing additional unstable elements because they are also physically and chemically stable. To further enhance the operational stability of PSCs, using inorganic layers with more vital bonding abilities to perovskite surfaces might be a viable approach. A unique passivation process was published by Yang et al. in 2019. Using an in situ reaction with certain inorganic anions, this method produced a thin, compact coating of inorganic lead sulfate ($PbSO_4$) on the perovskite surface. This strategy involved immersing the perovskite films in solutions comprising $SO_4{}^{2-}$ ions, which then reacted with the perovskite to form the $PbSO_4$ layer (Yang et al. 2019).

$$APbX_3 + SO_4^{2-} \rightarrow A^+ + 3X^- + PbSO_4 \tag{3.1}$$

The reaction time and precursor concentration had been optimized to ensure the formation of a thin layer (less than 5 nm) that would effectively passivate the Pb^{2+} imperfections without affecting the bulk perovskite frameworks. Additionally, strong SO_4^{2-} and Pb^{2+} ionic bonds also provided effective protection against degradation in humid conditions (Eq. 3.1). The resultant encapsulated PSCs preserved 96.8% of PCE after 1200 h of testing under standard irradiation in air with around 60 ± 10% RH and around 65 °C, while untreated PSCs only retained 46.9% of PCE after 474 h under identical conditions. The findings suggest that this inorganic passivation strategy could be a promising strategy for further enhancing the operational stability of PSCs.

Inorganic graphene oxide (GO) layers have shown potential for optoelectronic applications due to their remarkable electrical, optical, and physical characteristics and the low-cost solution-phase manufacturing methods. However, their weak interaction with perovskites has limited their application in enhancing the operational stability of PSCs. To address this issue, Wang et al. produced a Pb-rich surface on the perovskite film by using a diluted $Pb(SCN)_2$. Subsequently, the perovskite surface was coated with chlorinated graphene oxide (Cl-GO), which resulted in the formation of two robust Pb–O and Pb–Cl chemical bonds. The Pb–O bond within perovskite/Cl-GO is more durable than that inside perovskite/GO due to O's improved electron-withdrawing properties in Cl-GO. Once strong interaction between Cl-GO and the perovskite surface was developed, the encapsulated PSC tested at MPP under standard irradiance for 200 h at 60 °C demonstrated a considerable drop in the count of the I^- signal from the HTL layer. The encapsulated PSCs showed 90% PCE retention after 1000 h of MPP tracking at 60 °C and 1 solar irradiation as a result of all the previously described advances (Wang et al. 2019d).

In summary, the benefits of utilizing organic layers as passivation for the perovskite surface include (i) direct interaction with uncoordinated atoms to passivate surface defects, (ii) multi-functionality through the presence of amphiphilic groups, and (iii) the development of an outermost layer that effectively prevents ions from moving across the devices and prevents the direct contact with moisture, etc. Discovering and advancing such functional organic layers is essential to attain more reliable PSCs under practical operational conditions. In addition, inorganic layers offer three distinct benefits in addition to defect passivation when compared to organic layers: (i) perovskites and inorganic compounds have the potential to generate stronger ionic connections, as opposed to the weaker coordination bonds formed between organic groups and perovskites, resulting in more stable passivation. (ii) Compared to organic molecules, inorganic molecules have better mechanical and chemical stability, making them an effective protective layer for perovskite films against environmental factors like water and oxygen. (iii) The movement of ions from perovskites to charge transport layers has been effectively inhibited by inorganic layers, due to a decrease in the interactions between various layers. It is necessary to investigate these advantages further in order to obtain more stable PSCs under practical operating settings, as they all contribute to the passivation impact of inorganic layers on perovskites.

3.4.3.2 Structural Dimensionality Modification

In 3D perovskites, the low-dimensional (LD) materials are typically organic layers. Despite this, they exhibit unique properties such as photoactivity and wider bandgaps compared to 3D perovskites. These properties help to reduce the interfacial recombination of free carriers, and due to their shared inorganic component and framework with 3D perovskites, they can help to eliminate more interface defects by in situ interaction with LD components. Moreover, 3D perovskites can be shielded from water-induced degradation by the hydrophobic carbon rings or chains found in LD components. Taking everything into account, these benefits increase PSCs' operating stability in challenging environmental conditions. 2D perovskites are the most widely used LD material; they are also widely employed to improve PSCs' stability and efficacy against light, humidity, and heat stress (Lin et al. 2018a).

Grancini and colleagues (2017) used $(HOOC(CH_2)_4NH_3)_2PbI_4$ and $MAPbI_3$ to form an extraordinarily stable junction between 2 and 3D perovskites. The final structure included a systematically multi-dimensional interface, and after 10,000 h of MPP tracking under standard conditions, the PSCs generated from this junction maintained 100% of their initial PCE. The PSCs remained stable even after being subjected to temperature cycling ranging from 35 to 90 °C. The findings suggest a novel method for enhancing the operational stability of PSCs in humid conditions. By forming hydrophobic 2D layers, ligand vapor is applied to the PSC devices, repairing interface defects in 3D perovskites. To produce extremely effective and stable PSCs, more research should be done on the novel post-treatment strategy for the complete device (Grancini et al. 2017).

In 2019, Yoo et al. investigated the use of n-butyl-ammonium bromide, n-octyl-ammonium bromide, and n-hexyl-ammonium bromide as the 2D layer materials to passivate the surface and grain boundary (GB) defects of 3D perovskites (Fig. 3.10a). The passivation effects were similar among the three carbon chains, and the respective PSCs had better optoelectronic performance than untreated devices. The results demonstrated that the extent of the carbon chain had minimal impact on the passivation effect on the 3D perovskite layer. For 500 h of MPP tracking in ambient environments, a comparable PSC maintained 85% of its original efficiency due to the longer carbon chain (C_8Br) having greater hydrophobic characteristics (Fig. 3.10b, c), due to improved stability (Yoo et al. 2019).

Liu et al. described a method for creating a pure 2D perovskite film on the surface of 3D perovskites by submerging them in an isopropanol solution containing pentafluorophenylethylammonium iodide (FEAI). The reaction is produced by forming a 2D perovskite film composed of FEA_2PbI_4 on top of the 3D perovskite layer. This

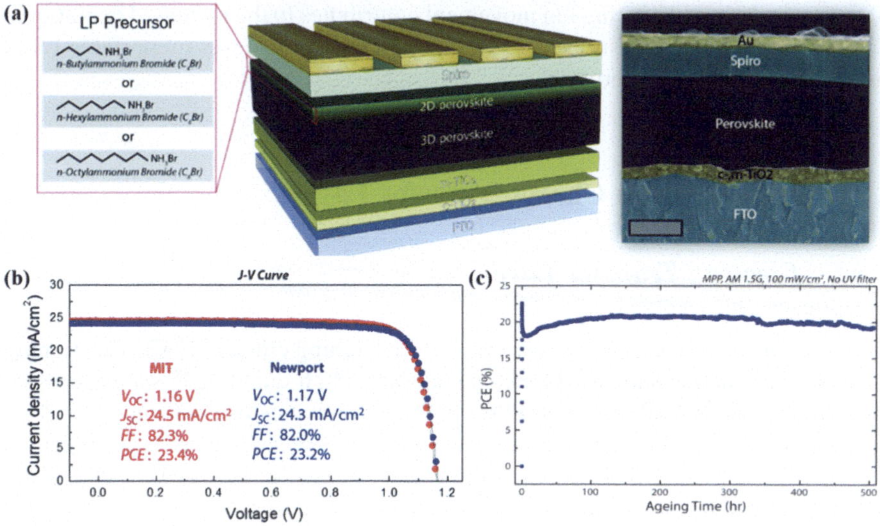

Fig. 3.10 **a** Schematics of a 3D/LP PSC and its cross-sectional image, **b** J–V analysis, and **c** stability test under MPP tracing at AM1.5G standard illumination without any UV filter. Reproduced from Ref. Yoo et al. (2019), Copyright (2019) with permission of the Royal Society of Chemistry

reduced the non-perovskite phase in the 3D perovskite and improved carrier transportation while decreasing interfacial recombination. The resulting 3D/2D bilayer PSC showed better functional stability than the pure 3D PSC, maintaining 90% of PCE under MPP tracking for 1000 h. The pure 3D PSC, in comparison, had a PCE drop to 43% after 1000 h under the same conditions. When Pyrrolidinium hydroiodide (PyI) salts were added to the surface of $FAPbI_3$ perovskites, Xu et al. observed that a 1D perovskite structure ($PyPbI_3$) formed on top of the underlying 3D perovskite. They carried out a comprehensive analysis and discovered that the formation of the 1D/3D structure reduced the activation barrier for temperature-dependent phase transition, greatly increasing the phase stability of $FAPbI_3$ PSCs. After 27 days in the air at a relative humidity of roughly 65%, the unencapsulated 1D/3D PSC devices retained 70% of PCE. In contrast, the control 3D devices lost their PCE after 12 days (Liu et al. 2019).

For eliminating perovskite defects, different molecules can be used for surface passivation, as summarized in Table 3.3. Surface passivators have several advantages over additives used in perovskite films. Because of post-treatment, they do not significantly affect the bulk properties of perovskites, which allows for the use of various molecules without consideration of their potential adverse effects on perovskite film formation. Furthermore, surface passivation can be either directly hydrophobic inorganic compounds or organic molecules with hydrophobic groups, which increases the stability of PSCs in humid environments. Moreover, by interacting with the surface ions of perovskite or creating a steric hindrance that prevents ions like I^- from moving freely inside the perovskite, surface passivation can function as an effective ion barrier. Limiting ion movement contributes to the increased operational stability of PSCs. Hence, surface passivation effectively achieves more stable PSCs under actual operating conditions.

3.5 Engineering on Charge Transfer Layers

3.5.1 Electron Transfer Layer

Electron transport layers (ETLs) are crucial in designing efficient PSCs, functioning as electron extraction and hole-blocking materials. High-quality ETLs are necessary for achieving decent efficiency in PSCs in the current state of the technology. Two main categories of ETL materials were suggested as potential candidates: fullerene derivatives such as PCBM and C_{60}, and metal oxides such as TiO_2, SnO_2, ZnO, and $BaSnO_3$. Metal oxides offer the best optical transparency across the electrical conductivity and absorber spectrum; however in order to guarantee the basic crystal structure needed to realize the optoelectronic properties, they must be treated at temperatures higher than 400 °C. Such high temperatures are incompatible with current device architectures and can cause damage to other functional layers and the absorber. The instability factors of metal oxide-based ETLs are due to the lower crystallinity resulting from the absence of high-temperature synthesis and potential photoinstability. Furthermore, a quick trap-prompted recombination at the TiO_2 interface could result from photoinduced oxygen desorption on TiO_2 surfaces. In contrast, fullerene derivatives are unstable under light illumination and can cause electronic disorders to trap carriers, leading to deterioration of device performance. Thus, the main challenge for achieving the operational stability of PSCs is the light-induced degradation of ETLs. The upcoming section will provide an overview of different ETLs and their effective modification approaches for achieving operational stability in PSCs (Dkhili et al. 2022; Mahmood et al. 2017).

3.5.1.1 Metal Oxides

TiO_2 was the first ETL used in PSCs, and many high-performing PSCs still rely on mesoporous TiO_2 films that are sintered at high temperatures (450–550 °C). However, the high-temperature processing of mesoporous TiO_2 films makes manufacturing more complicated and hinders the fabrication of bendable modules and tandem devices based on perovskites. Additionally, the photocatalytic properties of TiO_2 cause oxygen vacancies to develop in the presence of light, which causes trap-induced recombination to occur quickly across the TiO_2 interface. Consequently, mesoporous TiO_2 ETL is not favorable for the industrialization of PSCs. To address this problem, Zhou et al. reported that a planar TiO_2 made at a low temperature (150 °C) had desirable electronic characteristics. By doping the TiO_2 film with Y, its

Table 3.3 Typical summary of lifetime testing under different operational conditions and surface passivation

Device structure	Surface passivator	Test parameters	PCE retention	References
ITO/SnO_2/ $(FAPbI_3)_x(MAPbBr_3)_{1-x}$/PTAA/Au	$C_7H_8N_4O_2$	1-Sun illumination T: 40 °C, RH: 30–40%	500 h @95%	Wang et al. (2019a)
FTO/c-TiO_2/C_{60}/ $FA_{0.87}Cs_{0.13}Pb(I_{0.87}Br_{0.13})_3$/ spiro-MeOTAD/Au	$C_{2n}H_{4n+2}O_{n+1}$	MPP tracking under 1-Sun illumination RH: 50%	14 h @100%	Kim et al. (2018b)
ITO/SnO_2/$FA_xMA_{1-x}Pb_{1+y}I_3$/ PTAA/Au	(CCl)n	MPP tracking under 1-Sun illumination T: 60 °C	1000 h @90%	Wang et al. (2019d)
ITO/NiO_x/$MAPbI_3$/PCBM:C_{60}/ ZrAcac/Ag	$C_4H_{13}N_3$	MPP tracking under 1-Sun illumination T: 25 ± 1 °C RH: 50 ± 10%	500 h @70%	Zhang et al. (2018)
ITO/PTAA/ $Cs_{0.05}FA_{0.81}MA_{0.14}PbI_{2.55}Br_{0.45}$/ C_{60}/BCP/Cu	$PbSO_4$	MPP tracking under 1-Sun illumination T: 65 °C RH: 60 ± 10%	1200 h @96.8%	Yang et al. (2019)
FTO/c-TiO_2/m-TiO_2/$CsFAMAPbI_3$/ spiro-MeOTAD/Au	Pentafluorophenylethylammonium iodide	MPP tracking under 1-Sun illumination RH: 40%	1000 h @90%	Liu et al. (2019)

conductivity was greatly improved. Additionally, the ETL/ITO interface was effectively modified by inserting an extremely thin layer of PEIE to allow enhanced electron transfer ETL to ITO, which showed an increased PCE of 19.3% (Zhou et al. 2014). Chen et al. successfully developed large-area *p–i–n* PSCs using TiO_2 as the ETL. To avoid heat-induced degradation that would have resulted in amorphous TiO_2 and low conductivity, the annealing temperature of TiO_2 films was maintained below 70 °C. To address this, they partially substituted Ti^{4+} with Nb^{5+} in the TiO_2 film to heavily n-dope it, resulting in improved conductivity. After aging for 1000 h under one solar irradiation at 45–50 °C, the resulting large-sized PSCs retained 90% of their initial PCE and exhibited a high PCE of 16.2%. These findings suggest that PSCs using low-temperature processed planar TiO_2 or doped TiO_2 films as the ETL can also achieve greater PCEs and long-term operational stability, which is an encouraging development for satisfying market demands (Chen et al. 2015).

In order to mitigate TiO_2's photoinstability and enhance PSCs' long-term stability, substitute ETL materials such as SnO_2 and modified SnO_2 have been used. High-quality SnO_2 films were produced using a low-temperature, solution-based method developed by Anaraki et al., which were further improved by chemical bath deposition to enhance their hole-blocking ability. After 60 h of MPP tracking under standard conditions, the PSCs showing a PCE of 21% and retained 83.3% of it. This showed that improved performance and operational stability in PSCs were dependent on having a uniform, pinhole-free SnO_2 coating (Fig. 3.11a–c) (Anaraki et al. 2016). When EDTA was added to the SnO_2 film by Yang et al., the electron mobility of the E-SnO_2 film increased by around three times when compared to the untreated SnO_2 film.. This effectively eliminated charge accumulation at the perovskite/ETL interface. Planar-type PSCs produced as a result had a high PCE of 21.6%, surpassing the PCE attained with SnO_2 ETLs. Furthermore, after 120 h of aging under 1-Sun illumination, the E-SnO_2-based PSCs exhibited better long-term stability, maintaining 86% of their initial PCE, compared to the PSCs based on SnO_2, which only retained 38% of their initial efficiency (Yang et al. 2018a).

Because of its large 3.2 eV bandgap, $BaSnO_3$ (BSO), a kind of Sn-based oxide, has attracted attention as a possible n-type semiconducting perovskite. La-doped BSO (LBSO) is a desirable ETL for PSC devices due to its high electrical mobility at room temperature and negligible UV photocatalytic activity. However, LBSO typically requires a high crystallization temperature, which limits its applicability to flexible substrates. To address this issue, Shin et al. created a low-temperature processing that uses a crystalline superoxide-molecular cluster (CSMC) colloidal solution prepared of uniformly dispersed CSMC LBSO nanoparticles in 2-methoxyethanol. A dense and uniform LBSO layer was obtained by spin coating on the FTO substrate and annealing it through an intermediate peroxo-complex below 300 °C. After 1000 h of 1-Sun irradiation at 25 °C, the resultant PSCs, which included the low-temperature produced LBSO ETL, showed 21.2% PCE and preserved its 93.3%. This is in contrast to PSCs based on TiO_2, which fully degraded within 500 h under identical conditions (Shin et al. 2017).

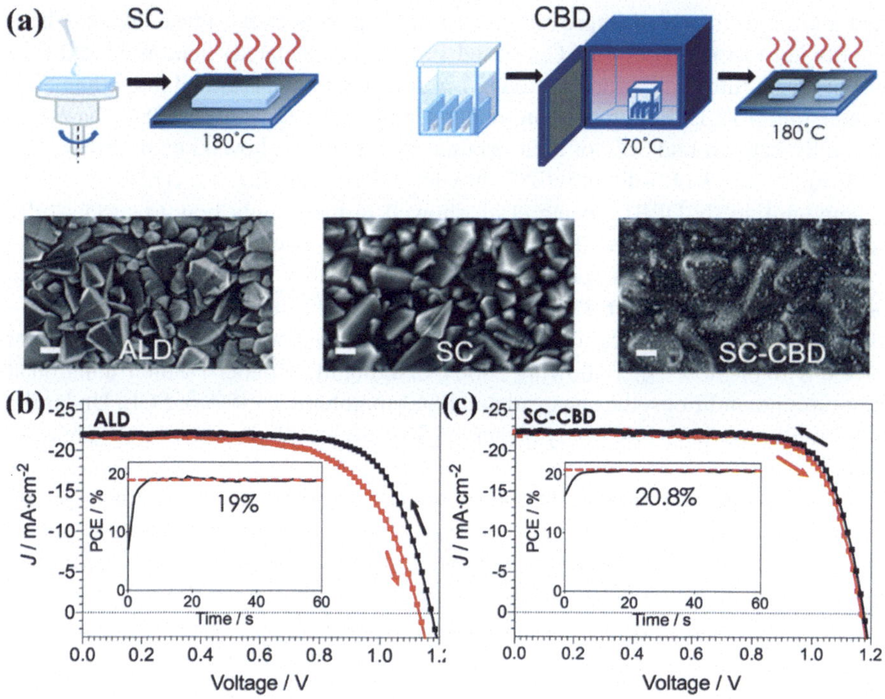

Fig. 3.11 **a** Schematics of SnO_2 film deposition using various vapor and solution processing methods, and **b**, **c** *J–V* analysis and stability of consequent PSCs, Reproduced from Ref. Anaraki et al. (2016), Copyright (2016) with permission of the Royal Society of Chemistry

Elevated temperatures are typically required to obtain highly crystalline metal oxide ETLs. Nonetheless, reducing the fabrication temperature without compromising the materials' optoelectronic features continues to be a difficult task. Optoelectronic properties of these materials remain a significant challenge. As discussed earlier, various approaches, such as ion doping and process modification, have been utilized to achieve efficient metal oxide ETLs at lower temperatures. These techniques offer promising strategies that can be further developed to broaden the range of applications for perovskite PV technology.

3.5.1.2 Fullerenes

Fullerenes, like PCBM and C_{60}, are frequently used as n-type materials in PSCs. In 2015, Wojciechowski and colleagues investigated the application of C_{60} as the ETL in conventional *n–i–p* topologies of PSCs. Because of the better band alignment and electrical interaction between the perovskite and C_{60}, they found that the perovskite was more efficient than TiO_2 at extracting electrons from C_{60}. Additionally, C_{60}'s ability to self-assemble made it easier for the bottom electrode and ETL to make chemical contact, which improved electron transport between these layers. These benefits led to the creation of C_{60}-based PSCs (encapsulated) that, after 500 h of MPP under 1-Sun irradiation (without a UV filter) at 60 °C, retained more than 50% of their initial PCE. In comparison, under identical circumstances, PSCs based on TiO_2 only kept around 30% of their original PCE (Wojciechowski et al. 2015).

Using 4-(1,3-dimethyl-2,3-dihydro-1H-benzimidazol-2-yl)-N, N-diphenylaniline (N-DPBI), Wang et al. showed in their study how to successfully n-doped C_{60}. It was noted that N-DPBI's aromatic phenyl rings were essential in maintaining the radical species by resonance, which produced a stable doped formation over time. After a thorough analysis, they concluded that the n-doping effect was generated by the C_{60} radical anions created when the N-DPBI dopant reacted with the host C_{60}. Following 2000 h of aging in air under 1-Sun irradiation at open-circuit conditions, the encapsulated PSCs maintained ~75% of PCE. Under the same conditions, bare-PSCs only preserved 56% after 2000 h of testing (Wang et al. 2017c).

The results thus far indicate that fullerenes have significant potential in achieving operationally stable and especially effective PSCs. Moreover, fullerenes' adherence to low-temperature processes and solubility in moderately polar solvents like chlorobenzene make them suitable for usage in *n–i–p* and *p–i–n* PSCs. These types of devices are more often used for different kinds of PSCs than metal oxides.

3.5.2 *Hole Transfer Layer*

Hole transport layers (HTLs) play a critical role in determining the photovoltaic performance of PSCs by serving as materials for hole extraction and electron blocking. Organic HTLs like spiro-MeOTAD, PTAA, P3HT, and EH44, as well as inorganic HTLs like NiO_x, CuSCN, CuI, and so on, are the two types of HTLs that are commonly employed. Imperfect morphology, which can provide shunting pathways during charge transfer, and low conductivity, which can lead to unintentional carrier transportation and ion migration at the perovskite/HTL/electrode interface, are among the limitations of both types of HTLs. Metal salts are incompatible with organic molecules, making it difficult to introduce them in the case of organic HTLs. Adding metal salts can create additional pinholes in organic HTLs that have already been formed, and the metal salt's hygroscopicity and chemical activity might cause unfavorable reactions with polar molecules like H_2O and O_2. Additionally, metal ions' mobility can cause organic film degradation by migrating across the entire PSC device during operation (Hasan et al. 2022).

3.5.2.1 Spiro-MeOTAD

Spiro-MeOTAD is the most commonly employed HTL in advanced *n–i–p* PSCs due to its appropriate HOMO energy level and remarkable hole extraction and transport potential. Yet, PSCs using spiro-MeOTAD as the HTL have exhibited poor thermal stability owing to its low Tg (121 °C), which further decreases upon adding Li-TFSI. Furthermore, pinholes in spiro-MeOTAD layers can act as ion-migration channels, which over time might deteriorate the optoelectronic characteristics and stability of the PSCs. Ono et al. developed a new method to design HTLs based on spiro-MeOTAD that is free of pinholes. Small defects in the film can cause unwanted ion migration and degradation in PSCs. The researchers found that using chlorobenzene as a solvent resulted in spiro-MeOTAD films with numerous pinholes and voids and substances containing iodine on the top electrode. On the other hand, employing chloroform as a solvent eliminated pinholes, possibly due to the increased polarity of the solvent. This new method is important for achieving better operational stability in PSCs because pinholes can cause unwanted ion migration and allow water and oxygen penetration in actual operation conditions (Yang et al. 2020).

To enhance the conductivity of spiro-MeOTAD materials, lithium-bis (trifluoromethane) sulfonimide (Li-TFSI) is often added as a dopant. Li-TFSI produces more pinholes, which results in doped spiro-MeOTAD film even though it is hydrophobic and capable of absorbing water from the surroundings. In order to overcome this issue, Spiro-MeOTAD(TFSI)$_2$, a potent p-dopant molecule substituted for Li-TFSI, was developed by Tan et al., described by the reaction given below.

$$\text{spiro - MeOTAD}^{2+} + \text{spiro - MeOTAD} \rightarrow 2\ \text{spiro - MeOTAD}^{+} \qquad (3.2)$$

It provides pure spiro-MeOTAD + devoid of contaminants such as Li + , which enhances the HTL films' conductivity by lowering series resistance, speeding up hole extractions, and minimizing interfacial charge trapping. Furthermore, the spiro-MeOTAD-(TFSI)$_2$-mixed HTLs displayed improved morphology, lacking the formation of aggregates or voids even in atmospheric conditions (Fig. 3.12). Consequently, 300 h after being exposed to standard illumination at a temperature of roughly 50 °C and relative humidity of roughly 40%, the resultant unencapsulated PSCs retained 65% of their original PCE. In contrast, the PSCs based on Li-TFSI retained 40% of their original PCE after 300 h under identical conditions. The findings suggest that the strategic modification of spiro-MeOTAD molecules can produce more durable HTLs, which is a promising area for future research and development (Tan et al. 2019).

3.5.2.2 Other Promising HTLs

Two types of novel HTLs have been developed: organic and inorganic materials. In 2018, for achieving electronic and charge transport features similar to spiro-MeOTAD, Christians and colleagues reported a Li$^+$-free hydrophobic HTL developed using EH44 (9-(2-ethylhexyl)-N, N, N, N-tetrakis(4-methoxyphenyl)-9H-carbazole-2,7-diamine) that contains π-conjugated methoxyphenyl groups (Fig. 3.12e). Additionally, the carbazole group in EH44 enables hydrophobic alkyl functionalization

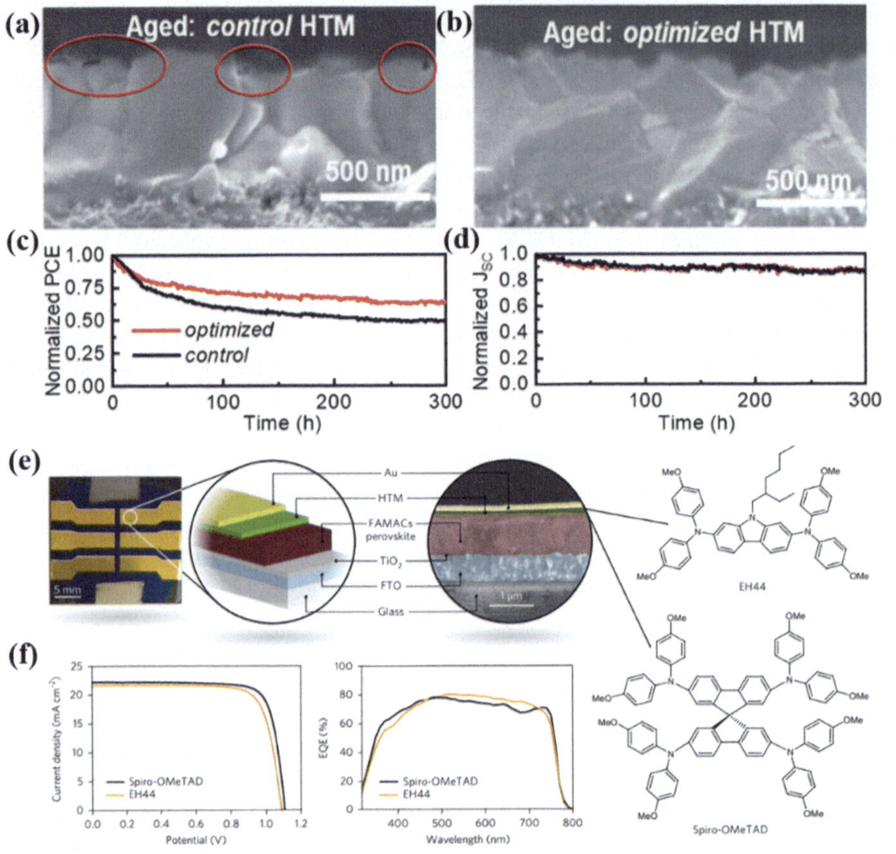

Fig. 3.12 **a**, **b** Cross-sectional FESEM image of fresh and aged optimized PSC device, **c** normalized PCE, **d** normalized J_{SC} showing good stability of 300 h hours under 1-Sun illumination at a temperature of around 50 °C and RH of about 40%, Reproduced from Ref. Tan et al. (2019), Copyright (2019) with permission of WILEY-VCH Verlag GmbH & Co. KGaA, Weinheim. **e**, **f** Schematics of general PSC architecture TiO_2/FAMACs/HTM/Au and their performance analysis, Reproduced from Ref. Christians et al. (2018), Copyright (2018) with permission of Springer Nature

at the N position. On employing silver bis(trifluoromethanesulfonyl)imide (Ag-TFSI) to oxidize EH44, $EH44^{+}TFSI^{-}$ is produced, which improves conductivity. The produced PSCs displayed a PCE of 18.5%, which is comparable to that of the device built using spiro-MeOTAD (19.6%) (Fig. 3.12f) (Christians et al. 2018).

Poor electrical coupling of P3HT molecules and perovskites, as well as inadequate hole transportation from the perovskite to P3HT, have also been linked to reduced PCE, according to prior studies. Jung et al. devised a double-layered halide architecture (DHA) in response to these issues. In this DHA, the HTL was deposited following the stacking of an ultrathin, wide-bandgap halide layer above a narrow-bandgap halide light-absorbing layer. This result was caused by the n-hexyl trimethyl ammonium bromide (HTAB) in situ reaction on the perovskite surface. Subsequent analysis revealed that P3HT's alkyl chains interdigitated with HTAB's ($C_6H_{13}^-$) similar alkyl chains facilitated enhanced van der Waals interactions and accelerated P3HT's promoted self-assembly upon deposit on the DHA surface. Consequently, the resulting PSCs' photovoltaic performance showed improvement in optoelectronic properties, with a certified PCE of 22.7%. Furthermore, even 1370 h later, these PSCs maintained 95% of their initial PCE, demonstrating improved operational stability. Additional materials that possess higher thermal stability, like PTAA, MPA-BTTI, and DBC-MeODPA, have also been demonstrated to enhance the operational stability of PSCs (Wang et al. 2019b; Ren et al. 2019).

Inorganic materials like NiO, CuI, and CuSCN are promising for achieving stable PSCs due to their higher thermal stability and lower fabrication cost than organic HTLs. However, a significant challenge with using these materials in solution processing is that the solvents used to dissolve them can lead to deterioration of the perovskite layer. Arora et al. used a quick drop-casting technique to apply a CuSCN solution of diethyl sulfide onto the perovskite layer to prevent degradation. The study showed that CuSCN domains had uniform coverage and were aligned parallel to the substrate's long unit cell axis. Additionally, by utilizing reduced graphene oxide (rGO) as a spacer between the HTL and Au contacts, the undesired resistance that formed between gold and the thiocyanate anions as a consequence of an electric field-induced reaction was avoided. Even after 1000 h of MPP tracking with these adjustments and without encapsulation, the PSCs maintained > 95% of their initial efficiency and reached a peak PCE of over 20% (Arora et al. 2017).

NiO_x has become a popular HTL material in PSCs due to its excellent stability. Several studies have demonstrated its effectiveness in enhancing the operational stability of PSCs under real-world operating conditions. Collectively, these findings suggest that inorganic HTLs hold great promise for developing stable PSCs.

3.6 Device Encapsulation

Encapsulation is a widely used technique to improve the functional stability of solar cells, particularly PSCs, and other electrical devices. Most notably, encapsulation serves as a barrier to reduce the loss of volatile compounds from the decomposition of perovskite; it also prevents the entrance of moisture and oxygen. Encapsulation has multiple benefits in strengthening perovskite devices. Various encapsulation methods have been attempted, focusing on the encapsulating materials that may significantly affect the perovskite layer. Using EVA to prevent perovskite oxidation and moisture infiltration, low-temperature processed PIB as encapsulants in a glass-glass encapsulation technique, and thermally cross-linkable ER films to encapsulate in double glass encapsulation for hail impact stability depending on the natural healing characteristics of ER films are a few examples of these techniques. These encapsulation techniques strengthen PSCs' resistance to moisture, thermal stress, and hail impact. It is currently uncertain whether the process of encapsulating PSCs has any effect on their long-term stability when subjected to operational conditions (Wang et al. 2021).

Ma et al. introduced a new encapsulation approach in 2020 that uses paraffin as the encapsulation material for PSCs, which is solvent-free and can be carried out at low temperatures. Compared to traditional encapsulation methods that use UV-curable adhesive, which may emit vapor during UV-curing that can harm the perovskite, this new method has a greater affinity with the delicate perovskite absorber. Furthermore, the paraffin's non-polarity helps eliminate any moisture and oxygen that may have remained after the encapsulation treatment, which prevents the perovskite from phase-segregating and prevents volatilization, which is the process by which it breaks down. The use of paraffin as an encapsulation material enhanced the thermal and moisture stability of PSCs, as demonstrated in several literature. Furthermore, this method of encapsulation, which is solvent-free, can be carried out in a typical environment and is, therefore, appropriate for mass production. This provides a new direction for advancing encapsulation technology for the industrialization of perovskite optoelectronics (Ma et al. 2020).

In addition to the encapsulation procedure, "functionalized encapsulation" entails enhancing the stability and environmental friendliness of PSCs by adding functional layers to either their front or rear side. A fluoropolymeric layer with luminous down-shifting (LDS) is applied to the glass side of PSCs to mitigate the harmful effects of UV radiation, and a fluoropolymeric layer with great moisture resistance is applied to the contact side of PSCs to help stabilize their humidity. Bella et al. showed that encapsulated PSCs could retain 95% of their original efficiency when operating outdoors by combining these two layers.

Device encapsulation must be used in the future to guarantee the practical application of PSCs. Advanced encapsulation technologies that include solvent-free processes and the introduction of functional layers can aid in producing more stable and environment-friendly PSCs that can operate effectively in different environments. A summary of the relevant modifications that can be improved further to enhance PSC commercialization is shown in Table 3.4.

3.7 Standard ISOS Protocol for PSCs Operational Stability

From the above discussion, it can be inferred that several external conditions, including outside environment, temperature, and illumination source intensity and type, influence the operational lifespan of PSCs (Howard et al. 2018; Holzhey et al. 2019). However, due to the diverse test methods and conditions, there is a need for more directly comparable results related to device stability. To address this issue, it is imperative to develop a uniform stability test standard that can generate reliable data, facilitating an appropriate interpretation and evaluation by the community. Organic photovoltaics (OPV) offer a useful reference, as they have already set up standardized aging experiments under the International Summit on Organic PV Stability (ISOS). The ISOS has established a comprehensive set of test protocols for evaluating the

Table 3.4 Typical summary of lifetime testing under different operational conditions and encapsulation

Device structure	Encapsulation	Testing parameters	PCE retention	References
ITO/SnO_2/RbCsMAFAPb$(I_{1-x}Br_x)_3$/ spiro-MeOTAD/Au	Paraffin	MPP tracking, 1-Sun illumination RH: 30–50%	1000 h @80%	Ma et al. (2020)
FTO/c-TiO_2/m-TiO_2/ $FA_{1-x}MA_xPbI_{3-y}Br_y$/spiro-MeOTAD/ Au	Luminescent down conversion material	Open environment under operational conditions	2160 h @95%	Bella et al. (2016)
ITO/NiO_x/$MAPbI_3$/PCBM/AZO/ITO	NiO_x as HTL	MPP tracking, 1-Sun illumination At 30 °C	4000 h @100%	Islam et al. (2019)
FTO/TiO_2/ $(CsPbI_3)_{0.05}(FAPbI_3)_{0.85}(MAPbBr_3)_{0.15}$/ spiro-MeOTAD/Au	Pb-active layer	MPP tracking, 1-Sun illumination At 50 °C	500 h @80%	Li et al. (2020)

stability of organic photovoltaics, which can also be applied to PSCs. The three tiers of these protocols—basic, intermediate, and advanced—cover a variety of testing scenarios, such as thermal cycling, simulated light and stress testing, dark, outdoor, and more. The parameters for each protocol include temperature, humidity, light, and electrical load, and the levels of stringency and complexity increase with each level (Table 3.5). Level 1 is the simplest, requiring basic equipment and setup requirements like dark storage, while levels 2 and 3 are more rigorous, including MPP tracking and providing more reliable results (Reese et al. 2011).

To develop appropriate testing procedures, Domansk et al. therefore carefully investigated the effects of varied operating conditions on the degrading behavior of PSCs. They compared the degradation behavior under open-circuit, short-circuit, and MPP conditions (Fig. 3.13a–e) and noticed that aging under MPP conditions was the best option for assessing PSCs stability as it simulates real practical working conditions. Aging at open or short circuits resulted in conservative estimations. Based on their results, they suggested that the most reliable way to assess operational stability for aging PSCs is to use MPP tracking under standard condition at both room and high temperatures. Additionally, they noticed that the deterioration curve typically included two components: a linear decline and an exponential decline. It was possible to reverse the reversible process of exponential decay and restore the device's functionality by keeping it in the dark for a long time (Domanski et al. 2018).

Table 3.5 General overview of various ISOS protocols for device testing under different operational conditions

Test type	Dark			Light		
Test ID	ISOS-D-1	ISOS-D-2	ISOS-D-3	ISOS-O-1	ISOS-O-2	ISOS-O-3
Storage	Shelf	High temp. storage	Damp heat	Outdoor Condition	Outdoor Condition	Outdoor Condition
Temp	RT	RT	RT	Sunlight	Sunlight	Sunlight
RH	Ambient	65–85 °C	65–85 °C	Ambient	Ambient	Ambient
Light source	Ambient	Oven	Env. Chamber	Sunlight	Sunlight	Sunlight
Load	OC	OC	OC	MPP	MPP	MPP

	Laboratory weathering testing			Thermal cycling testing		
	ISOS-L-1	ISOS-L-2	ISOS-L-3	ISOS-T-1	ISOS-T-2	ISOS-T-3
Temp	Ambient	65/85 °C	Simulator	–	–	–
RH	Ambient	Ambient	65–85 °C	65–85 °C	65–85 °C	-40 to + 85 °C
Setup	Light only	Light and Temp.	Ambient	Ambient	Near 50%	Near 55%
Light source	Solar simulator	Solar simulator	Light, temp. and RH	Hot plate/ oven	Oven/Env. Chamber	Env. chamber
Load	Open circuit (OC)	Open circuit	MPP	Solar simulator (OC)	Solar simulator (OC)	Solar simulator (OC)

	Solar-thermal-humidity cycling		
	ISOS-LT-1	ISOS-LT-2	ISOS-LT-3
Temp	RT to 65 °C	5 to 65 °C	– 25 and 65 °C
RH	Uncontrolled	Controlled (50%)	Controlled (50%)
Setup	Weathering chamber	Env. chamber and sun simulation	Env. chamber and sun simulation
Light source	Solar simulator	Solar simulator	Solar simulator
Load	MPP	MPP	MPP

Reproduced from Ref. Reese et al. (2011), Copyright (2011) with permission of Elsevier

Scientists at the 11th International Summit on Organic and Hybrid Photovoltaics Stability discussed the suitability of ISOS stress methods for PSCs and recommended additional stress testing. They also highlighted the need for more transparency and repeatability in reporting scientific and procedural details regarding the stability testing of PSCs. Recently, Khenkin et al. compiled and published pertinent meeting contents, including MPP test procedures (listed in Table 3.6). This emphasizes how crucial it is to carry out and honestly report on intricate experimental

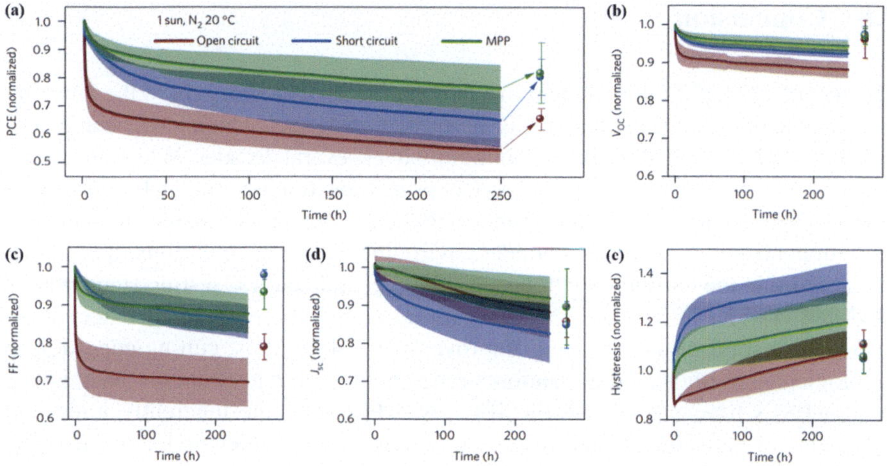

Fig. 3.13 Comparison of the degradation behavior under open-circuit, short-circuit, and MPP conditions. **a–e** PCE, V_{OC}, fill factor (FF), J_{SC}, and $J–V$ hysteresis-evolution, Reproduced from Ref. Domanski et al. (2018), Copyright (2018) with permission of Springer Nature

procedures in order to produce reliable MPP data, including initial solar cell characterization, encapsulation, aging conditions, aging period, measurements throughout aging, sample count, and outdoor stability (Khenkin et al. 2020).

Table 3.6 Compiled and published relevant contents of the device testing, including MPP test procedures and results that one looks at during device testing

Important points	Features	Points to be looked on
Initial characterizations	*J–V* analysis, MPP tracking, IPCE analysis	A light source, dwelling time, scan speed, spectrum, intensity, tracking algorithm, delay time, tracking duration, etc.
Aging conditions	Light, temperature, environment, bias parameters, cycling processes	Light intensity, spectrum, filters, dark/illumination, air/glovebox/sealed/Env. chamber/RH, open circuit/MPP/load, dwelling time, etc.
Outdoor stability	Environment, exposure, and weather	Coordinates, total duration of exposure, temperature, RH, solar irradiance, etc.
Aging measurements	Periodic *J–V* analysis, MPP tracking	Frequency recording, scan parameters, tracking parameters, recovery time, etc.
Sealing	Wiring, encapsulation, Edge sealant	Materials, processing parameters, active area, reference, sealing thickness, rim, environmental conditions, etc.

3.8 Conclusion

The recent and rapid improvement in PSCs can be attributed mostly to the amazing and tunable optoelectronic features of perovskite materials, which have been extensively studied by the PV research community over several decades. While efficiency and cost are no longer significant concerns in transitioning from lab research to commercialization of PSCs, the primary challenge lies in addressing their limited operational lifetime, particularly under environmental aging conditions. It has been suggested that the current comprehensive and generalized IEC testing standards be modified in order to evaluate PSC stability in a range of environmental scenarios, providing a solid basis. By using improved perovskite films, cutting-edge device topologies, and efficient stabilization strategies through encapsulation techniques, certain PSCs have passed specific IEC tests. However, the inherently ionic and soft nature of perovskite materials presents challenges in achieving long-term operational stability due to phenomena such as ion migration and phase segregation induced by illumination and electric fields. The development of operationally stable PSCs is of utmost importance for advancing the commercialization of perovskite PV technologies.

Furthermore, it is crucial to address the availability and reproducibility of stability results for PSCs by establishing a standardized testing process for their operational stability. Additionally, future publications should include detailed experimental procedures to ensure the reliability and comparability of results. The goal of creating a logical testing and evaluation standard and improving the functional stability of PSCs is becoming more and more important to the research community. Similar to the rapid advancements in PSC efficiency, it is anticipated that operational stability will also improve rapidly through efficient and effective research endeavors, ultimately facilitating the commercialization of PSCs.

References

Abdi-Jalebi M, Andaji-Garmaroudi Z, Cacovich S et al (2018) Maximizing and stabilizing luminescence from halide perovskites with potassium passivation. Nature 555:497–501. https://doi.org/10.1038/nature25989

Aharon S, Layani M, El CB et al (2015) Self-assembly of perovskite for fabrication of semitransparent perovskite solar cells. Adv Mater Interfaces 2:1–6. https://doi.org/10.1002/admi.201500118

Anaraki EH, Kermanpur A, Steier L et al (2016) Highly efficient and stable planar perovskite solar cells by solution-processed tin oxide. Energy Environ Sci 9:3128–3134. https://doi.org/10.1039/c6ee02390h

Ansari MIH, Qurashi A, Nazeeruddin MK (2018) Frontiers, opportunities, and challenges in perovskite solar cells: a critical review. J Photochem Photobiol C Photochem Rev 35:1–24. https://doi.org/10.1016/j.jphotochemrev.2017.11.002

Arora N, Dar MI, Hinderhofer A et al (2017) Perovskite solar cells with CuSCN hole extraction layers yield stabilized efficiencies greater than 20%. Science (80-) 358:768–771. https://doi.org/10.1126/science.aam5655

Azpiroz JM, Mosconi E, Bisquert J, De Angelis F (2015) Defect migration in methylammonium lead iodide and its role in perovskite solar cell operation. Energy Environ Sci 8:2118–2127. https://doi.org/10.1039/c5ee01265a

Bella F, Griffini G, Correa-Baena JP et al (2016) Improving efficiency and stability of perovskite solar cells with photocurable fluoropolymers. Science (80-) 354:203–206. https://doi.org/10.1126/science.aah4046

Brivio F, Caetano C, Walsh A (2016) Thermodynamic origin of photoinstability in the $CH_3NH_3Pb(I_{1-x}Br_x)_3$ hybrid halide perovskite alloy. J Phys Chem Lett 7:1083–1087. https://doi.org/10.1021/acs.jpclett.6b00226

Bush KA, Palmstrom AF, Yu ZJ et al (2017) 23.6%-efficient monolithic perovskite/silicon tandem solar cells with improved stability. Nat Energy 2:1–7. https://doi.org/10.1038/nenergy.2017.9

Cappel UB, Svanström S, Lanzilotto V et al (2017) Partially reversible photoinduced chemical changes in a mixed-ion perovskite material for solar cells. ACS Appl Mater Interfaces 9:34970–34978. https://doi.org/10.1021/acsami.7b10643

Chen W, Wu Y, Yue Y et al (2015) Efficient and stable large-area perovskite solar cells with inorganic charge extraction layers. Science (80-) 350:944–948. https://doi.org/10.1126/science.aad1015

Chen B, Rudd PN, Yang S et al (2019a) Imperfections and their passivation in halide perovskite solar cells. Chem Soc Rev 48:3842–3867. https://doi.org/10.1039/c8cs00853a

Chen Y, Li N, Wang L et al (2019b) Impacts of alkaline on the defects property and crystallization kinetics in perovskite solar cells. Nat Commun 10. https://doi.org/10.1038/s41467-019-09093-1

Christians JA, Schulz P, Tinkham JS et al (2018) Tailored interfaces of unencapsulated perovskite solar cells for >1000 hour operational stability. Nat Energy 3:68–74. https://doi.org/10.1038/s41560-017-0067-y

De Angelis F et al (2017) Perovskite solar cells on their way to the market. ACS Energy Lett 2:2640–2641. https://doi.org/10.1021/acsenergylett.7b01038

Dkhili M, Lucarelli G, De Rossi F et al (2022) Attributes of high-performance electron transport layers for perovskite solar cells on flexible PET versus on glass. ACS Appl Energy Mater 5:4096–4107. https://doi.org/10.1021/acsaem.1c03311

Domanski K, Alharbi EA, Hagfeldt A et al (2018) Systematic investigation of the impact of operation conditions on the degradation behaviour of perovskite solar cells. Nat Energy 3:61–67. https://doi.org/10.1038/s41560-017-0060-5

Eames C, Frost JM, Barnes PRF et al (2015) Ionic transport in hybrid lead iodide perovskite solar cells. Nat Commun 6:2–9. https://doi.org/10.1038/ncomms8497

El-Mellouhi F, Marzouk A, Bentria ET et al (2016) Hydrogen bonding and stability of hybrid organic-inorganic perovskites. Chemsuschem 9:2648–2655. https://doi.org/10.1002/cssc.201600864

Grancini G, Roldán-Carmona C, Zimmermann I et al (2017) One-year stable perovskite solar cells by 2D/3D interface engineering. Nat Commun 8:1–8. https://doi.org/10.1038/ncomms15684

Gu F, Ye S, Zhao Z et al (2018) Improving performance of lead-free formamidinium tin triiodide perovskite solar cells by tin source purification. Sol RRL 2:1–9. https://doi.org/10.1002/solr.201800136

Gui D, Ji L, Muhammad A et al (2018) Jahn-Teller effect on framework flexibility of hybrid organic-inorganic perovskites. J Phys Chem Lett 9:751–755. https://doi.org/10.1021/acs.jpclett.7b03229

Hasan MS, Alom J, Asaduzzaman M et al (2022) Recent criterion on stability enhancement of perovskite solar cells. Processes 10. https://doi.org/10.3390/pr10071408

Hoke ET, Slotcavage DJ, Dohner ER et al (2015) Reversible photo-induced trap formation in mixed-halide hybrid perovskites for photovoltaics. Chem Sci 6:613–617. https://doi.org/10.1039/c4sc03141e

Holzhey P, Saliba M et al (2018) A full overview of international standards assessing the long-term stability of perovskite solar cells. J Mater Chem A 6:21794–21808. https://doi.org/10.1039/C8TA06950F

Holzhey P, Yadav P, Turren-Cruz SH et al (2019) A chain is as strong as its weakest link—stability study of $MAPbI_3$ under light and temperature. Mater Today 29:10–19. https://doi.org/10.1016/j.mattod.2018.10.017

Howard JM, Tennyson EM, Barik S et al (2018) Humidity-induced photoluminescence hysteresis in variable Cs/Br ratio hybrid perovskites. J Phys Chem Lett 9:3463–3469. https://doi.org/10.1021/acs.jpclett.8b01357

IEC W (2016) IEC 61215-1:2016

Islam MB, Yanagida M, Shirai Y et al (2019) Highly stable semi-transparent $MAPbI_3$ perovskite solar cells with operational output for 4000 h. Sol Energy Mater Sol Cells 195:323–329. https://doi.org/10.1016/j.solmat.2019.03.004

Jacobsson T, Jesper J-PC-B, Pazoki M, Saliba M et al (2016) An exploration of the compositional space for mixed lead halogen perovskites for high efficiency solar cells. Energy Environ Sci 9:1706–1724

Jiang Y, Leyden MR, Qiu L et al (2018) Combination of hybrid CVD and cation exchange for upscaling Cs-substituted mixed cation perovskite solar cells with high efficiency and stability. Adv Funct Mater 28. https://doi.org/10.1002/adfm.201703835

Jiang Y, Qiu L, Juarez-perez EJ et al (2019a) halide perovskite solar modules using self-healing polymer-based encapsulation. Nat Energy 4. https://doi.org/10.1038/s41560-019-0406-2

Jiang Y, Qiu L, Juarez-Perez EJ et al (2019b) Reduction of lead leakage from damaged lead halide perovskite solar modules using self-healing polymer-based encapsulation. Nat Energy 4:585–593. https://doi.org/10.1038/s41560-019-0406-2

Khenkin MV, Katz EA, Abate A et al (2020) Consensus statement for stability assessment and reporting for perovskite photovoltaics based on ISOS procedures. Nat Energy 5:35–49. https://doi.org/10.1038/s41560-019-0529-5

Kim HS, Seo JY, Park NG (2016) Material and device stability in perovskite solar cells. Chemsuschem 9:2528–2540. https://doi.org/10.1002/cssc.201600915

Kim J, Park N, Yun JS et al (2017) An effective method of predicting perovskite solar cell lifetime–Case study on planar $CH_3NH_3PbI_3$ and $HC(NH_2)_2PbI_3$ perovskite solar cells and hole transfer materials of spiro-OMeTAD and PTAA. Sol Energy Mater Sol Cells 162:41–46. https://doi.org/10.1016/j.solmat.2016.12.043

Kim GY, Senocrate A, Yang TY et al (2018a) Large tunable photoeffect on ion conduction in halide perovskites and implications for photodecomposition. Nat Mater 17:445–449. https://doi.org/10.1038/s41563-018-0038-0

Kim M, Motti SG, Sorrentino R, Petrozza A (2018b) Enhanced solar cell stability by hygroscopic polymer passivation of metal halide perovskite thin film. Energy Environ Sci 11:2609–2619. https://doi.org/10.1039/c8ee01101j

Koh TM, Fu K, Fang Y et al (2014) Formamidinium-containing metal-halide: an alternative material for near-IR absorption perovskite solar cells. J Phys Chem C 118:16458–16462. https://doi.org/10.1021/jp411112k

Kour R, Arya S, Verma S et al (2019) Potential substitutes for replacement of lead in perovskite solar cells: a review. Glob Challenges 3:1900050. https://doi.org/10.1002/gch2.201900050

Lee S, Kang DW et al (2017) Highly efficient and stable Sn-rich perovskite solar cells by introducing bromine. ACS Appl Mater Interfaces 9:22432–22439. https://doi.org/10.1021/acsami.7b04011

Li N, Tao S, Chen Y et al (2019) Cation and anion immobilization through chemical bonding enhancement with fluorides for stable halide perovskite solar cells. Nat Energy 4:408–415. https://doi.org/10.1038/s41560-019-0382-6

Li X, Zhang F, He H et al (2020) On-device lead sequestration for perovskite solar cells. Nature 578:555–558. https://doi.org/10.1038/s41586-020-2001-x

Lin Y, Bai Y, Fang Y et al (2018a) Enhanced thermal stability in perovskite solar cells by assembling 2d/3d stacking structures. J Phys Chem Lett 9:654–658. https://doi.org/10.1021/acs.jpclett.7b02679

Lin Y, Chen B, Fang Y et al (2018b) Excess charge-carrier induced instability of hybrid perovskites. Nat Commun 9:1–9. https://doi.org/10.1038/s41467-018-07438-w

Liu Z, Hu J, Jiao H et al (2017) Chemical reduction of intrinsic defects in thicker heterojunction planar perovskite solar cells. Adv Mater 29:1–8. https://doi.org/10.1002/adma.201606774

Liu L, Huang S, Lu Y et al (2018) Grain-boundary "patches" by in situ conversion to enhance perovskite solar cells stability. Adv Mater 30:1–8. https://doi.org/10.1002/adma.201800544

Liu Y, Akin S, Pan L et al (2019) Ultrahydrophobic 3D/2D fluoroarene bilayer-based water-resistant perovskite solar cells with efficiencies exceeding 22%. Sci Adv 5:1–9. https://doi.org/10.1126/sciadv.aaw2543

Ma S, Bai Y, Wang H et al (2020) 1000 h operational lifetime perovskite solar cells by ambient melting encapsulation. Adv Energy Mater 10:1–8. https://doi.org/10.1002/aenm.201902472

Mahmood K, Sarwar S, Mehran MT (2017) Current status of electron transport layers in perovskite solar cells: materials and properties. RSC Adv 7:17044–17062. https://doi.org/10.1039/c7ra00002b

Min H, Kim M, Lee SU et al (2019) Efficient, stable solar cells by using inherent bandgap of a-phase formamidinium lead iodide. Science (80-) 366:749–753. https://doi.org/10.1126/science.aay7044

Nie R, Sumukam RR, Reddy SH et al (2020) Lead-free perovskite solar cells enabled by hetero-valent substitutes. Energy Environ Sci 13:2363–2385. https://doi.org/10.1039/d0ee01153c

Park B, Seok SI (2019) Intrinsic instability of inorganic–organic hybrid halide perovskite materials. Adv Mater 31:1–17. https://doi.org/10.1002/adma.201805337

Ran C, Xu J, Gao W et al (2018) Defects in metal triiodide perovskite materials towards high-performance solar cells: origin, impact, characterization, and engineering. Chem Soc Rev 47:4581–4610. https://doi.org/10.1039/c7cs00868f

Reese MO, Gevorgyan SA, Jørgensen M et al (2011) Consensus stability testing protocols for organic photovoltaic materials and devices. Sol Energy Mater Sol Cells 95:1253–1267. https://doi.org/10.1016/j.solmat.2011.01.036

Ren M, Wang J, Xie X et al (2019) Double-helicene-based hole-transporter for perovskite solar cells with 22% efficiency and operation durability. ACS Energy Lett 4:2683–2688. https://doi.org/10.1021/acsenergylett.9b01949

Rolston N, Bennett-Kennett R, Schelhas LT et al (2020) Comment on "light-induced lattice expansion leads to high-efficiency perovskite solar cells." Science (80-) 368:67–70. https://doi.org/10.1126/science.aay8691

Saliba M, Matsui T, Seo JY et al (2016) Cesium-containing triple cation perovskite solar cells: improved stability, reproducibility and high efficiency. Energy Environ Sci 9:1989–1997. https://doi.org/10.1039/c5ee03874j

Sherkar TS, Momblona C, Gil-Escrig L et al (2017) Recombination in perovskite solar cells: significance of grain boundaries, interface traps, and defect ions. ACS Energy Lett 2:1214–1222. https://doi.org/10.1021/acsenergylett.7b00236

Shi L, Young TL, Kim J et al (2017) Accelerated lifetime testing of organic-inorganic perovskite solar cells encapsulated by polyisobutylene. ACS Appl Mater Interfaces 9:25073–25081. https://doi.org/10.1021/acsami.7b07625

Shin SS, Yeom EJ, Yang WS et al (2017) Colloidally prepared La-doped $BaSnO_3$ electrodes for efficient, photostable perovskite solar cells. Science (80-) 356:167–171. https://doi.org/10.1126/science.aam6620

Tan B, Raga SR, Chesman ASR et al (2019) LiTFSI-free spiro-OMeTAD-based perovskite solar cells with power conversion efficiencies exceeding 19%. Adv Energy Mater 9:1–10. https://doi.org/10.1002/aenm.201901519

Tavakoli MM, Saliba M, Yadav P et al (2019) Synergistic crystal and interface engineering for efficient and stable perovskite photovoltaics. Adv Energy Mater 9:1–8. https://doi.org/10.1002/aenm.201802646

Tong J, Song Z, Kim DH et al (2019) Carrier lifetimes of >1 ms in Sn-Pb perovskites enable efficient all-perovskite tandem solar cells. Science (80-) 364:475–479. https://doi.org/10.1126/science.aav7911

Wang Q, Chen B, Liu Y et al (2017a) Scaling behavior of moisture-induced grain degradation in polycrystalline hybrid perovskite thin films. Energy Environ Sci 10:516–522. https://doi.org/10.1039/c6ee02941h

Wang Z, Lin Q, Chmiel FP et al (2017b) Efficient ambient-air-stable solar cells with 2D–3D heterostructured butylammonium-caesium-formamidinium lead halide perovskites. Nat Energy 2:1–10. https://doi.org/10.1038/nenergy.2017.135

Wang Z, McMeekin DP, Sakai N et al (2017c) Efficient and air-stable mixed-cation lead mixed-halide perovskite solar cells with n-doped organic electron extraction layers. Adv Mater 29:1–8. https://doi.org/10.1002/adma.201604186

Wang R, Xue J, Wang KL et al (2019a) Constructive molecular configurations for surface-defect passivation of perovskite photovoltaics. Science (80-) 366:1509–1513. https://doi.org/10.1126/science.aay9698

Wang Y, Chen W, Wang L et al (2019b) Dopant-free small-molecule hole-transporting material for inverted perovskite solar cells with efficiency exceeding 21%. Adv Mater 31:1–10. https://doi.org/10.1002/adma.201902781

Wang Y, Ibrahim Dar M, Ono LK et al (2019c) Thermodynamically stabilized β-$CsPbI_3$–based perovskite solar cells with efficiencies >18%. Science (80-) 365:591–595. https://doi.org/10.1126/science.aav8680

Wang Y, Wu T, Barbaud J et al (2019d) Stabilizing heterostructures of soft perovskite semiconductors. Science (80-) 365:687–691. https://doi.org/10.1126/science.aax8018

Wang Y, Ahmad I, Leung T et al (2021) Encapsulation and stability testing of perovskite solar cells for real life applications. ACS Mater Au. https://doi.org/10.1021/acsmaterialsau.1c00045

Wojciechowski K, Leijtens T, Siprova S et al (2015) C60 as an efficient n-type compact layer in perovskite solar cells. J Phys Chem Lett 6:2399–2405. https://doi.org/10.1021/acs.jpclett.5b00902

Wu W, Yang Z, Rudd PN et al (2019) Bilateral alkylamine for suppressing charge recombination and improving stability in blade-coated perovskite solar cells. 1–10

Xu J, Buin A, Ip AH et al (2015) Perovskite-fullerene hybrid materials suppress hysteresis in planar diodes. Nat Commun 6:1–8. https://doi.org/10.1038/ncomms8081

Yang D, Yang R, Wang K et al (2018a) High efficiency planar-type perovskite solar cells with negligible hysteresis using EDTA-complexed SnO_2. Nat Commun 9. https://doi.org/10.1038/s41467-018-05760-x

Yang TCJ, Fiala P, Jeangros Q, Ballif C (2018b) High-bandgap perovskite materials for multijunction solar cells. Joule 2:1421–1436. https://doi.org/10.1016/j.joule.2018.05.008

Yang S, Chen S, Mosconi E et al (2019) Stabilizing halide perovskite surfaces for solar cell operation with wide-bandgap lead oxysalts. Science (80-) 365:473–478. https://doi.org/10.1126/science.aax3294

Yang Y, Hoang MT, Yao D et al (2020) Spiro-OMeTAD or CuSCN as a preferable hole transport material for carbon-based planar perovskite solar cells. J Mater Chem A 8:12723–12734. https://doi.org/10.1039/d0ta03951a

Yin WJ, Shi T, Yan Y (2014) Unusual defect physics in $CH_3NH_3PbI_3$ perovskite solar cell absorber. Appl Phys Lett 104. https://doi.org/10.1063/1.4864778

Yoo JJ, Wieghold S, Sponseller MC et al (2019) An interface stabilized perovskite solar cell with high stabilized efficiency and low voltage loss. Energy Environ Sci 12:2192–2199. https://doi.org/10.1039/c9ee00751b

Zhang F, Shi W, Luo J et al (2017) Isomer-pure Bis-PCBM-assisted crystal engineering of perovskite solar cells showing excellent efficiency and stability. Adv Mater 29:1–7. https://doi.org/10.1002/adma.201606806

Zhang H, Ren X, Chen X et al (2018) Improving the stability and performance of perovskite solar cells: via off-the-shelf post-device ligand treatment. Energy Environ Sci 11:2253–2262. https://doi.org/10.1039/c8ee00580j

Zhang F, Zhu K et al (2020) Additive engineering for efficient and stable perovskite solar cells. Adv Energy Mater 10:1–26. https://doi.org/10.1002/aenm.201902579

Zheng X et al (2023) Strain relaxation for perovskite lattice reconfiguration. Adv Energy Sustain Res 4:2200143. https://doi.org/10.1002/aesr.202200143

Zhou H et al (2014) Interface engineering of highly efficient perovskite solar cells. Science (80-) 345:542–546

Zhu H, Liu Y, Eickemeyer FT et al (2020) Tailored Amphiphilic Molecular Mitigators for Stable Perovskite Solar Cells with 23.5% Efficiency. Adv Mater 32:1–8. https://doi.org/10.1002/adma.201907757

Chapter 4
Possible Perovskite Substitutes, Advances, and Future Prospects

Abstract Tin-based perovskites, particularly Sn-iodide ($ASnI_3$), exhibit a notable power conversion efficiency (PCE) of 14.6%, making them promising lead-free alternatives for solar cell applications. Despite challenges in stability due to rapid Sn^{2+} oxidation, recent chemical engineering advancements achieved over 1300 h of operational stability in a nitrogen environment. Sn/Ge mixed perovskite absorbers show potential but require efficiency enhancements. Notably, Sn-iodide perovskites with low bandgaps could serve as foundation materials for tandem solar cells. Among alkaline-earth metals, magnesium iodide perovskites offer a tunable visible bandgap (0.9 to 1.7 eV), while calcium, strontium, and barium halide perovskites face limitations with high bandgaps (2.95 to 3.6 eV) and humidity sensitivity. Metal chalcogenide perovskite semiconductors, designed through split-anion or anion alloying, present a non-toxic alternative with the potential for efficient solar energy conversion. Theoretical calculations indicate favorable bandgaps and increased optical absorption in the visible range, highlighting the potential of metal chalcogenide perovskite semiconductors for advancing photovoltaic technologies.

Keywords Lead-free metal halide perovskites (MHPs) · Alkaline-earth MHPs · Metal chalcogenide MHPs · Transition metal MHPs · Heterovalent MHPs · Lanthanide/actinide MHPs · Group 14 MHPs

4.1 Introduction

In the last ten years, there has been significant interest in halide perovskite semiconductors due to their ability to have adjustable bandgaps, excellent optoelectronic properties, simple processing, and inexpensive manufacturing costs, opening up a wider spectrum of possibilities for device applications. These semiconductors have been particularly useful in developing solar cells, light-emitting diodes, photocatalysts, and photodetectors. Promising materials for these applications include 3D hybrid halide inorganic perovskites made of lead, halogen anions, and organic/

A. Srivastava and P. M. Shirage, *Perovskite Solar Cells*, Springer Tracts in Electrical and Electronics Engineering, https://doi.org/10.1007/978-981-97-7624-5_4

inorganic cations. The first perovskites that had been effectively used as photovoltaic materials in a solar cell were $(CH_3NH_3{:}MA)PbI_3$ and $MAPbBr_3$ in 2009. When UV photoelectron or optical spectroscopy is used, $MAPbI_3$ has shown excellent features for applications in solar cells, including a strong optical absorption having an intense absorption coefficient of 10^4 cm^{-1} and a bandgap of 1.5–1.6 eV. Additionally, $MAPbI_3$ single crystals have sizeable charge carrier diffusion lengths varying between 3 and 10 μm and ultra-low trap densities between 10^9 to 10^{10} traps per cm^3 (Chen et al. 2015; Ibn-Mohammed et al. 2017; Manser et al. 2016).

The properties of Pb-based halide perovskites can be modified by replacing methylammonium (MA^+) with various cations like cesium (Cs^+), potassium (K^+), or formamidinium (FA^+). This substitution can alter the materials' thermodynamic, structural, and electronic properties. For instance, the stability of the materials can be increased, and the bandgap decreases by 0.07 eV by using a mixture of FA^+ and Cs^+. Mixing different halogen atoms to produce mixed-halide perovskites is another effective way to tune the bandgap and extend the absorption and emission energies across the UV–visible to near-infrared regions. Hence, finely tuned 3D halide perovskites composed of Pb-based alloys of cations and halogens are ideal for developing high-performance solar cells that include tandems on silicon. Currently, the maximum certified PCE a Pb-based perovskite solar cell has ever attained is 25.8%. This record had been recently set by inserting a coherent interlayer made of Cl^- between the $FAPbI_3$ and SnO_2 electrodes (Sani et al. 2018; Yang et al. 2019).

Although Pb-based halide perovskites exhibit excellent performance, their stability under atmospheric conditions and the ecological problems raised by Pb are severe obstacles to their widespread commercialization. Several approaches have been established to vanquish these obstacles. For example, encapsulating the perovskite materials within devices or modules can protect them against contact with moisture, UV light, heating, or air, mitigating their rapid degradation and enhancing their long-term stability. Alternatively, researchers are exploring Pb-free perovskites and perovskite-like structures (perovskitoids) that offer comparable or possibly superior properties compared to traditional Pb-based materials.

The Goldschmidt tolerance factor has been applied to hybrid organic–inorganic metal halide perovskite materials incorporating organic molecular cations like MA^+. This approach, which has been applied for predicting new lead-free perovskite compounds for photovoltaic applications, takes into account the ionic radii of the pertinent ions. The B-site and other ion locations in the perovskite structure can be identified using this method as potential replacement possibilities. The methodology has been demonstrated by predicting hundreds of hypothetical perovskites that could be used as potential candidates for di-valent replacement of lead with divalent metal cations that include alkaline-earth metal and lanthanide-based materials (Li et al. 2022; Travis et al. 2016). The concept of the tolerance factor has also been applied in various other investigations for predicting novel metal halide perovskites. Perovskite-like materials with less dimensionality than traditional perovskites can be developed outside of the stability range of the Goldschmidt tolerance factor. For instance, two-dimensional layered perovskites with a structure similar to Ruddlesden–Popper phases can be achieved by adding large A-site cations in the interlayer space. An example of this is $(MA)_2CuCl_xBr_{4-x}$. In contrast, for lower dimensional perovskite derivatives, which include one-dimensional chain-like or zero-dimensional structures, the Goldschmidt tolerance factor concept cannot be applied similarly due to the gradual lifting of the ionic size restrictions. One example of a lower dimensional perovskite-like material is $(MA)_3Sb_2I_9$ (Bartel et al. 2019).

4.2 Distinguished Lead-Free Perovskites

The reference perovskite structure is the cubic ABX_3 with a $pm\overline{3}m$ space group. This structure consists of BX_6 octahedra, which are corner-shared, while the A-site atom is located in the cuboctahedral cavity created between the octahedra, as depicted in Fig. 4.1. Typically, the A-site is commonly occupied by a monovalent organic or inorganic cation, the B-site by a divalent metal, such as Pb, and the X-site by the halogen anion. Despite the fact that A-site atoms only indirectly affect the electronic structure of these materials, they can cause structural modifications leading to changes in electronic structure and bandgap. Recently, based on the 3D and 2D ordered structures developed from the ABX_3 lattice, Akkerman and Manna classified various structure types of materials (King et al. 2010). Here, we focus on substitutionally engineered materials at the B-site, which replace all Pb atoms and have been used in optoelectronic devices to date. There are two general approaches for substituting lead with environmentally friendly elements in the formation of lead-free metal halide perovskites such as:

1. One of the approaches is di-valent substitution. This approach involves substituting lead with cations that have 2+ valency, like group-14 elements (Ge, Sn), alkaline-earth metals (Mg, Ca, Sr, Ba), transition metals (Mn, Fe, Ni, Pd, Cu, Cd), and lanthanides (Eu, Tm, Yb) (Kour et al. 2019).
2. Mono- and tri-valent substitution involves replacing lead with aliovalent cations like transition metals, main group elements, lanthanides, and actinides. Since these cations cannot provide charge neutrality in the ABX_3 structure, a mixed-valence approach can be used, in which an equal amount of monovalent and trivalent metal cations are added in order to balance the overall charge and valence. Double halide perovskites that are based on a mixture of various hetero-valent metal cations and $A_3B_2X_9$-type perovskites, based on a mixture of higher valent metal cations as well as vacancies, are also possible approaches for hetero-valent substitution. Despite this, the Goldschmidt replacement rules cannot be used to predict these replacement methods (Kour et al. 2019).

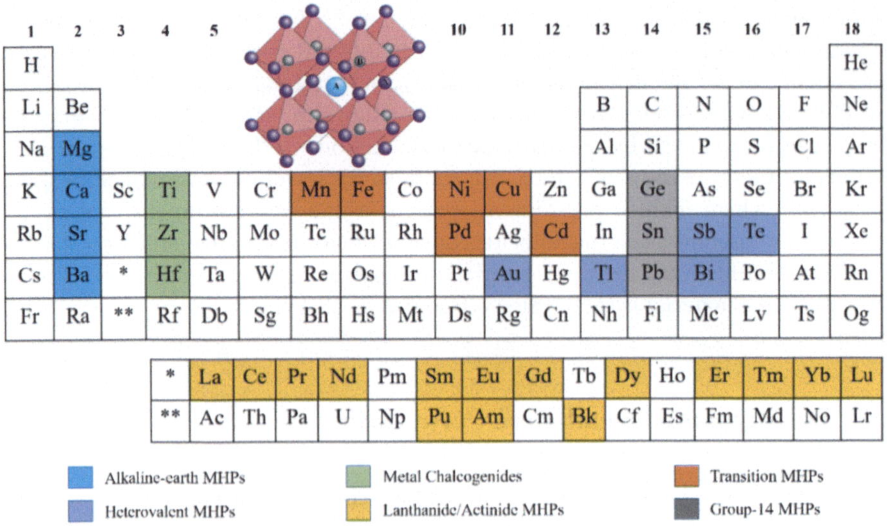

Fig. 4.1 Typical schematic of perovskite structure and diverse range of elements used to prepare lead-free metal halide perovskite semiconductors

As illustrated in Fig. 4.1, a diverse range of lead-free metal halide perovskite semiconductors can be made utilizing different elements from the periodic table by using di-, mono-, and tri-valent substitution strategies. The following sections will discuss these perovskites in detail.

4.3 Di-valent Substitution

A variety of elements with a stable oxidation state of +2 can potentially replace lead in the perovskite structure via di-valent substitution. Among these are group-14 elements (Ge^{2+}, Sn^{2+}), alkaline-earth metals (Be^{2+}, Mg^{2+}, Ca^{2+}, Sr^{2+}, Ba^{2+}), transition metals (V^{2+}, Mn^{2+}, Fe^{2+}, Co^{2+}, Ni^{2+}, Pd^{2+}, Pt^{2+}, Cu^{2+}, Zn^{2+}, Cd^{2+}, Hg^{2+}), lanthanides (Eu^{2+}, Tm^{2+}, Yb^{2+}), as well as p-block elements (Ga^{2+}, In^{2+}). However, some of these options are unsuitable because they have a limited capacity to produce perovskites, have high bandgaps (Be^{2+}, Ca^{2+}, Sr^{2+}, and Ba^{2+}), are toxic (Cd^{2+} and Hg^{2+}), are radioactive, or are unstable in the +2-oxidation state. The best potential candidates for lead-free perovskites were identified through computational screening of di-valent substitution in cesium and methylammonium metal halide perovskites as Sn^{2+}, Ge^{2+}, Mg^{2+}, V^{2+}, Mn^{2+}, Ni^{2+}, Zn^{2+}, and Co^{2+} (Filip et al. 2016; Kieslich et al. 2015).

4.3.1 Group-14 Elements

Group-14 elements, specifically Sn and Ge, are appropriate candidates for di-valent substitution of lead in perovskites due to their iso-electronic configurations as Pb^{2+}. Even though tin and germanium halide perovskites possess enhanced optoelectronic properties, the two of them, Sn^{2+} and Ge^{2+} ions, are susceptible to oxidation to the + 4-oxidation state, which is a result of the lowered inert pair effect and is more evident for Ge^{2+} than for Sn^{2+}. Despite this, several research teams in the perovskite field are currently concentrating on enhancing the stability of Sn and Ge-based halide perovskites, and only some techniques for doing so have already been described (Lim et al. 2021; Zhang et al. 2018).

4.3.1.1 Tin (Sn^{2+})

The toxic and hazardous nature of lead presents a major obstacle to the widespread adoption of Pb-based perovskite solar cells (PSCs). Due to its solubility and toxicity, lead can easily enter the human body by inhalation and travel to the brain via blood or olfactory nerves. As a result, it can impair a number of bodily systems, including the hematological, renal, reproductive, and central neurological systems. Researchers are exploring appropriate substitutions or alternatives to Pb-based perovskite absorbers to address these hazardous concerns, including OIHPs with cations such as tin, germanium, bismuth, antimony, copper, and manganese. Sn-based perovskites, anticipated to replace Pb-based perovskites as a non-toxic substitute, are thought to be the most promising of all the alternatives. This is primarily because of their analogous chemical and optoelectronic characteristics, which fulfill the requirements for coordination, ionic size, and charge balance.

Sn-based perovskites possess similar or better optoelectronic properties in comparison to Pb-based perovskites, such as a narrow bandgap (ranging from 1.28 to 1.55 eV)(Eperon et al. 2014), low exciton binding energy (< 100 meV) (Cai et al. 2018), higher carrier mobility (between 10^2 and 10^3 cm^2 V^{-1} s^{-1}), long carrier diffusion length (ranging from 10^2 to 10^5 nm) (Alcocer et al. 2013), high absorption coefficient ($> 10^{-4}$ cm^{-1}), and high conductivity (Li et al. 2019). According to theoretical calculations, the PCE limit for Sn-based PSCs can reach around 30% (Fig. 4.2a, b) (Eperon et al. 2014; Stranks et al. 2013). The story of success has not yet been reproduced with Sn-based PSCs, and published PCEs are still significantly lower than those of Pb-based PSCs despite the expertise garnered from the development of Pb-based devices. This is mostly because Sn-based perovskites are poorly stable in the presence of a variety of factors, including moisture, oxygen, light irradiation, applied electric fields, thermal stressors, and defects in the perovskite film. The quick oxidation of Sn^{2+} to the more stable Sn^{4+} and the generation of Sn vacancies with small formation energy result in high amounts of self-p-doping in perovskite films and substantial recombination losses, significantly hindering the development of tin-based PSCs (Lee et al. 2016).

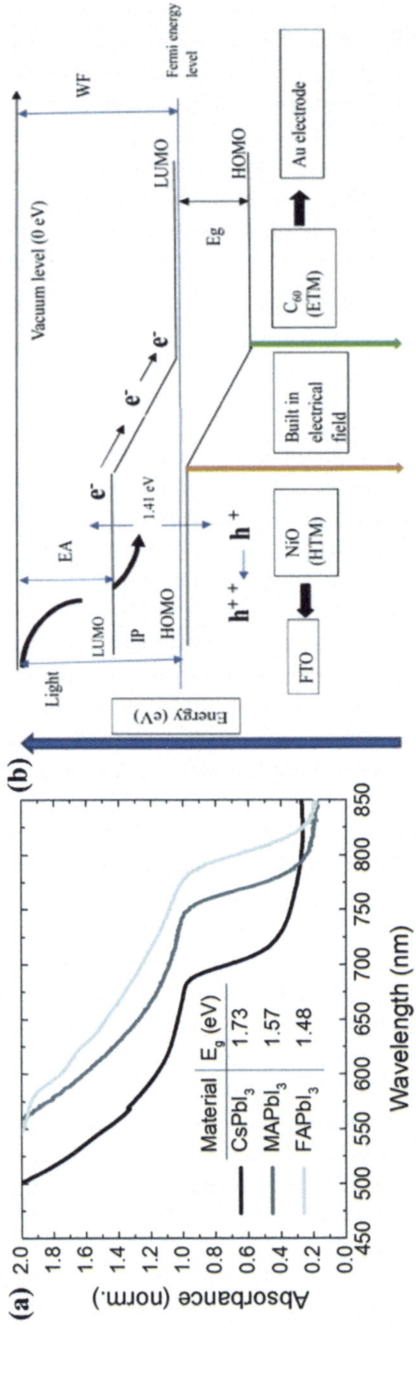

Fig. 4.2 **a** Typical bandgap of different Pb-based PSCs, Reproduced from Ref. Eperon et al. (2014), Copyright (2014) with permission of the Royal Society of Chemistry. **b** General band diagram for estimation theoretical PCE of Sn-based PSCs

The tin halide perovskites that have been extensively studied are $MASnI_3$ and $FASnI_3$. Like lead halide perovskites, tin-based perovskites' structural properties, such as the dimensionality and connectivity of the perovskite lattice, can be significantly influenced by the size, function, and halide used in the A-site cation. A-site cations that are small and monovalent, such as MA^+, FA^+, and Cs^+, result in the formation of 3-D structures, while larger ones like cyclobutylammonium and tropylium lead to reduced dimensionality, such as 2-D layered, 1-D chain-like, or 0-D structures. These changes in composition and structure also have an impact on the optical and electronic properties of the perovskites (Macdonald et al. 2022).

The morphology and compactness of Sn-based perovskite films are mainly determined by their rapid nucleation and growth kinetics. It is essential to control the crystallization kinetics during processing in order to enhance the morphology of the film. One effective strategy is to add an antioxidant such as tin-halide (SnF_2 or $SnCl_2$) into the solution. The inclusion of SnF_2 can reduce the concentration of Sn vacancies (V_{Sn}) by enhancing their formation energy, which creates more nucleation sites and promotes the growth of additional tin-based perovskite nuclei. As a result, the homogeneous crystal growth is facilitated, leading to a more uniform film coverage (Ayaydah et al. 2023). The use of DMSO is critical in order to achieve a uniform and dense morphology with fewer pinholes. DMSO forms a stable intermediate complex with FAI, DMSO, and SnI_2, called $SnI_2{\cdot}3DMSO$, which slows down the crystallization process and evenly distributes the precursor over the substrate during spin coating. Meanwhile, adding $SnCl_2$ and SnF_2 as antioxidants reduces the oxidation of Sn^{2+} and promotes the formation of nucleation sites that prevent dendritic growth, resulting in a more compact and pinhole-free film. During spin coating, anti-solvents are dripped onto the substrate to remove any surplus DMSO, creating a homogeneous precursor film that, upon annealing with the aid of heterogeneous SnF_2 nuclei, produces a uniform, dense film with equiaxed grains (Aftab and Ahmad 2020). Koh et al. conducted a study to investigate the impact of various amounts of SnF_2 (ranging from 0 to 40 mol%) on $FASnI_3$. Their results indicated that the presence of Sn^{4+} in $FASnI_3$ gradually decreased with increasing amounts of SnF_2, while the addition of 20% SnF_2 produced a more uniform and continuous film. The issue of $FASnI_3$ penetrating *m*-TiO_2 and forming a capping layer on top of the perovskite film, thereby hindering uniform coverage, was addressed by improving the thickness of *m*-TiO_2 up to 500 nm. This prevented capping and enabled $FASnI_3$ to fill the pores perfectly, increasing coverage and enhancing PCE. In particular, as depicted in Fig. 4.3a, increasing the *m*-TiO_2 thickness from 350 to 500 nm resulted in a rise in PCE from 1.71 to 2.1% (Koh et al. 2015).

Solvent engineering is a widely used technique to control the morphology and performance of PSCs. As shown in Fig. 4.3b, Zhao et al. employed chlorobenzene as an anti-solvent for mixed cation perovskite (FA, MA)SnI_3 films and observed improved film morphology, reduced charge carrier recombination and achieved an efficiency of 8.12% (Zhao et al. 2017) Liu et al. also employed anti-solvent dripping (diethyl ether, toluene, and chlorobenzene) along with SnF_2 addition in a mixture of FA and MA perovskite, resulting in efficiencies of 6.39%, 7.52%, and 9.06%, respectively, with negligible hysteresis as depicted in Fig. 4.3c (Liu et al. 2018). In

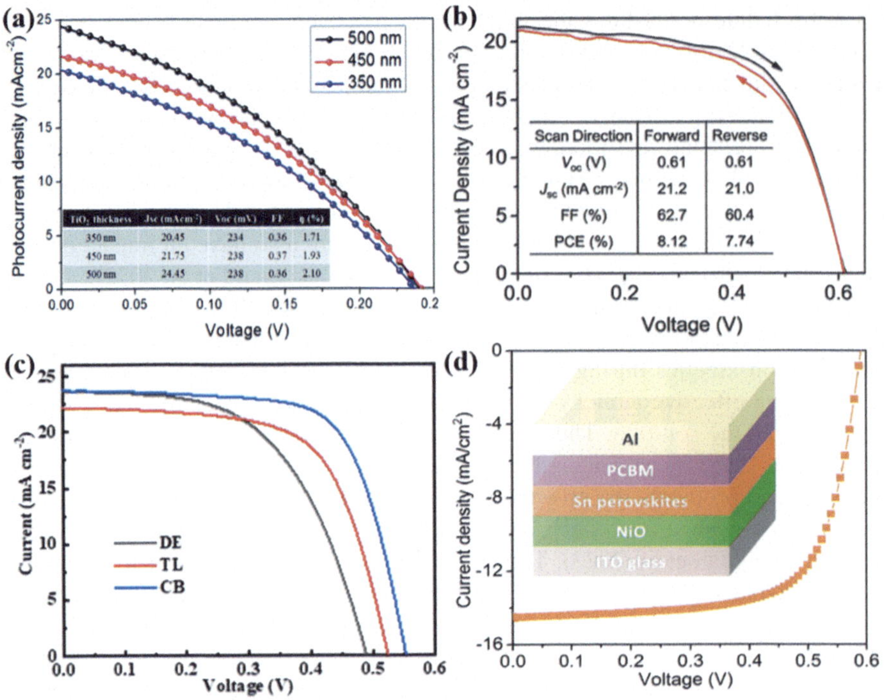

Fig. 4.3 **a** *J*–*V* analysis of SnF_2 treated $FASnI_3$ PSCs, Reproduced from Ref. Koh et al. (2015), Copyright (2015) with permission of the Royal Society of Chemistry. **b** Another example of PCE enhancement via treating mixed cation perovskite (MA, FA)SnI_3 PSC through chlorobenzene anti-solvent, Reproduced from Ref. Zhao et al. (2017), Copyright (2017) with permission of WILEY-VCH Verlag GmbH & Co. KGaA, Weinheim. **c** *J*–*V* analysis of anti-solvent (diethyl ether, toluene, and chlorobenzene) dripping mixed cation PSCs, Reproduced from Ref. Liu et al. (2018), Copyright (2018) with permission of the American Chemical Society. **d** *J*–*V* analysis of highly efficient $FASnI_3$ perovskite by incorporating 8% heavier cation PEA^+, Reproduced from Ref. Liao et al. (2017), Copyright (2017) with permission of the American Chemical Society

the beginning of 2018, Ran et al. reported the use of a bilateral interfacial engineering approach to develop effective lead-free Sn-based PSCs with a PCE of 6.98% (V_{OC} = 0.47 V, J_{SC} = 20.07 mA cm^{-2}, FF = 0.74) and better reproducibility. Shao et al. demonstrated another exceptionally reproducible Sn-based PSC with a PCE of 9% (V_{OC} = 0.52 V, J_{SC} = 24.1 mA cm^{-2}, FF = 0.71) by incorporating 8% heavier cation PEA^+ in $FASnI_3$, as depicted in Fig. 4.3d that enhanced the PSCs stability of the cell, retaining 59% of the initial PCE after storing for 76 h of storage (Liao et al. 2017). Similarly, Liao et al. investigated the effect of SnF_2 addition (0–30 mol%) combined with anti-solvent dripping (diethyl ether) on $FASnI_3$ films. They reported a compact morphology with uniform film coverage and an efficiency of 6.22% including 10% SnF_2 and diethyl ether dripping (Liao et al. 2016). Later, Wang et al. added a removable pseudo halogen regulator NH_4SCN to an HSP that was made of parallelly oriented 2D-quasi-2D-3D $FASnI_3$ and reported a PCE as high as 9.41% ($J_{SC} = 22.0$ mA cm^{-2}, $V_{OC} = 0.61$ V, FF = 0.70), with 90% of the PCE remaining after 600 h of storage in the glove box (Wang et al. 2018).

Jokar et al. investigated the effects of adding guanidinium (GA^+) in varying proportions to the formamidinium (FA^+) tin triiodide perovskite ($FASnI_3$) with an addition of 1% ethylene diammonium diodide ($EDAI_2$) as an additive. They attained a PCE of 9.6%, which is one of the best-recorded efficiencies for Sn-based PSCs ($J_{SC} = 21.2$ mA cm^{-2}, $V_{OC} = 0.619$ V, FF = 0.729). The stability of the cell was demonstrated without encapsulation in air for six days (Jokar et al. 2019). Jiang et al. developed a 2D/3D Sn-based perovskite solar cell by utilizing indene-C_{60} bis-adduct as an electron transporting material and incorporating SnF_2 into $PEA_{0.15}FA_{0.85}SnI_3$ perovskite. They employed toluene as an anti-solvent during the fabrication process and presented cross-sectional SEM images and J-V analysis, as shown in Fig. 4.4a, b. Their device exhibited an impressive PCE of 12.41%, presently the highest recorded efficiency (Jiang et al. 2020). The structural and optical characteristics of tin halide perovskites and how well they work as absorber materials in photovoltaic devices are summarized in Table 4.1.

Recent studies have identified several issues that hinder the stability and performance of Sn-based PSCs. The first issue is the tendency of tin to exhibit different oxidation states despite Sn^{4+} being more stable than Sn^{2+}. The Sn^{2+} state is more likely to oxidize to the Sn^{4+} state when exposed to oxygen and moisture because of its reduced stability. The second issue is the rapid crystallization of Sn-based perovskites, making it challenging to control how the microstructure evolves while processing, leading to the formation of non-uniform films comprising porosity and defects. As a result, there is a higher risk of device shorting, ineffective charge transport/collection, and charge recombination. The third issue is that the oxidation of Sn^{2+} produces Sn vacancy, which causes p-type self-doping, enhances leakage current and charge carrier recombination, limiting the open-circuit voltage ($V_{OC} \approx 0.5$ mV), and, therefore, lowers cell efficiency. Despite persistent efforts, the maximum PCE achieved so far is only ~10–12%.

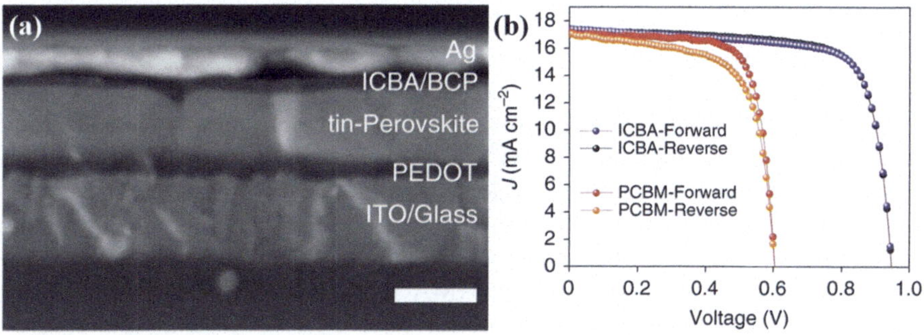

Fig. 4.4 **a, b** Cross-sectional image of ethyl toluene anti-solvent treated 2D/3D Sn-based perovskite solar cell consisting of indene-C_{60} bis-adduct as an electron transporting material and incorporating SnF_2 into $PEA_{0.15}FA_{0.85}SnI_3$ perovskite, and respective *J–V* analysis. Reproduced from Ref. Jiang et al. (2020), Copyright (2020) with permission of Springer Nature

Table 4.1 Typical structural and optical properties of various Sn-based halide perovskites and respective PCEs obtained with such Sn-based active layer PSCs

Crystal system	Space group	Absorber material	Bandgap (eV)	η (%)	References
Pseudocubic	P4mm	$MASnBr_3$	2.15–2.2	4.27	Hao et al. (2014)
		$MASnI_3$	1.27–1.35	5.23	
		$MASnIBr_2$	1.75	5.73	
		$MASnI_2Br$	1.56	5.48	
Orthorhombic	Pnma	$CsSnI_{2.95}F_{0.05}$	1.3	8.51	Chung et al. (2012)
	Amm2	$FASnI_3$	1.4–1.41	6.22	Liao et al. (2016)
	–	$FASnI_2Br$	1.68	1.72	Zhang et al. (2016a)
		$CsSnI_3$	1.27–1.31	3.31	Wang et al. (2016)
Cubic	Fm3m	$Cs_2SnI_3Br_3$	1.43	3.63	Kaltzoglou et al. (2015)
	Fm3mm	Cs_2SnCl_6	3.9	0.07	Kaltzoglou et al. (2016)
		Cs_2SnI_6	1.26–1.62	7.8	
		Cs_2SnBr_6	2.7	0.04	
	–	$CsSnIBr_2$	1.63–1.65	3.2	Li et al. (2016)

Sn perovskite faces challenges due to the poor stability of Sn^{2+} and uncontrolled crystallization during spin coating. Moisture, air, UV illumination, and temperature degrade the device, lowering efficiency. Hysteresis, charge recombination, and ion migration are additional problems. Strategies to improve stability and performance include controlling oxidation with additives, anti-solvent dripping, larger cation incorporation, Br anion substitution, and device encapsulation (Fig. 4.5) (Aftab et al. 2021).

4.3.1.2 Germanium (Ge^{2+})

Germanium is a promising alternative to lead for perovskite structures. Despite having the same oxidation state as Pb^{2+}, the divalent germanium cation (Ge^{2+}) is less electronegative, has a more covalent nature, and has a smaller ionic radius (73 pm) in comparison to Pb^{2+} (119 pm) (Chen et al. 2015). As demonstrated in the case of $MAGeX_3$ (X = Cl, Br, I) compounds with tolerance factor values of 1.005 ($MAGeCl_3$), 0.988 ($MAGeBr_3$), and 0.965 ($MAGeI_3$), that are consistent with t values reported for the ideal perovskite structure ($0.97 < t < 1.03$), the Goldschmidt tolerance factor calculations encourage the development of germanium halide perovskites (Sun et al. 2016a).

In 2015, Krishnamoorthy et al. attempted to incorporate Ge-based halide perovskites into photovoltaic devices. They screened various possible candidates

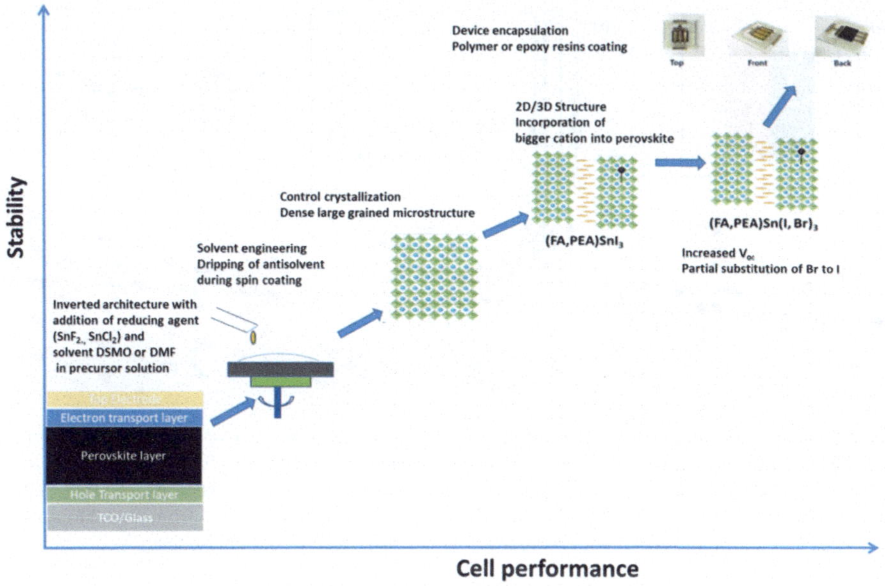

Fig. 4.5 Typical strategies to improve stability and performance include controlling oxidation with additives, anti-solvent dripping, larger cation incorporation, Br anion substitution, and device encapsulation. Reproduced from Ref. Aftab et al. (2021), Copyright (2021) with permission of Elsevier

and selected germanium iodide perovskites ($MAGeI_3$ and $CsGeI_3$) to fabricate solar cells. However, germanium perovskite dissolves poorly in the majority of organic polar solvents, and poor morphologies were produced, which had a detrimental effect on the performance of the device. The solar cells based on $MAGeI_3$ and $CsGeI_3$ achieved device PCEs of 0.11% and 0.20 accordingly, as shown in Fig. 4.6a, b (Krishnamoorthy et al. 2015). In recent times, Kopacic and colleagues examined the photovoltaic performance of $MAGeX_3$ (X = I, Br) and found that the perovskite became less susceptible to deterioration by mixing of halides, leading to improved solar cell performance. In another study, $MAGeI_3$ and $MAGeI_{2.7}Br_{0.3}$ were prepared using dimethylformamide as the solvent, and their stability under ambient conditions was analyzed using time-resolved UV–Vis spectroscopy. The mixed halide composition $MAGeI_{2.7}Br_{0.3}$ showed improved ambient stability and was found to be correlated with the lifetimes of the synthesized PSCs, which were better than those based on $MAGeI_3$ (Kopacic et al. 2018).

Advanced computational modeling has been employed alongside limited experimental research to investigate the electronic and optical properties of Ge-based perovskites. This has also helped in defining their stability and predicting potential novel compositions. Sun and colleagues conducted a systematic study of the properties of $MAGeX_3$ perovskites in the trigonal phase (R3m space group) by comparing them to Pb and Sn-based perovskites with tetragonal phase (I4/mcm). Detailed calculations were done on optical properties, electronic band structures,

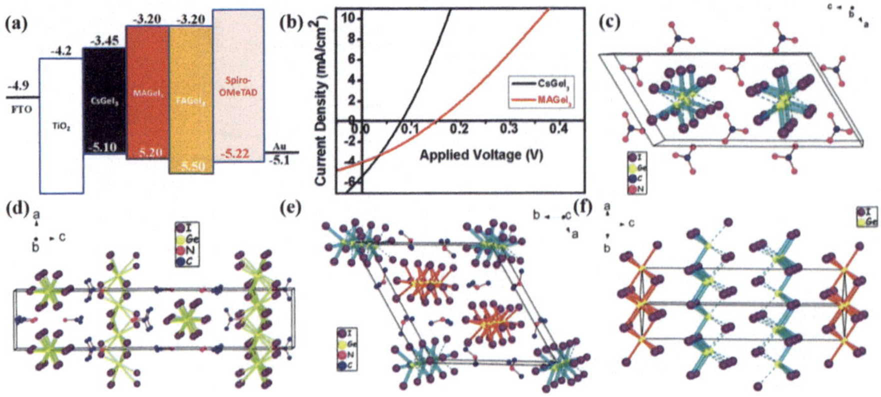

Fig. 4.6 **a** Typical band-structure of different Ge-based PSCs, **b** respective *J–V* analysis, Reproduced from Ref. Krishnamoorthy et al. (2015), Copyright (2015) with permission of Royal Society of Chemistry, **c–f** schematic crystal structures of 1- and 2-D Ge-iodides such as $C(NH_2)_3GeI_3$, $(CH_3)_3NHGeI_3$, $(CH_3)_2C(H)NH_3GeI_3$, and GeI_2, respectively, Reproduced from Ref. Stoumpos et al. (2015), Copyright (2015) with permission of the American Chemical Society

and structure and formation energies. The findings indicate that $MAGeI_3$ displays an electronic character similar to that of $MAPbI_3$. Ge-based perovskites are less susceptible to oxidation than Sn-based perovskites, resulting in a stability order of $MAPbI_3 > MAGeI_3 > MASnI_3$. The chemical instability of the divalent Ge^{2+} cation upon oxidation has limited the experimental investigation of germanium halide perovskites. This problem is more significant in germanium-based perovskites than tin-based perovskites because of the reduced inert electron pair effect (Chiara et al. 2021).

The structural, electronic, and optical properties of $AGeI_3$ germanium halide perovskites containing Cs^+ and various organic A-site cations were thoroughly investigated by Stoumpos et al. Based on the A-site cation used, various structures can be formed. Small cations like Cs^+, MA^+, or FA^+-ions facilitate the production of three-dimensional perovskite frameworks based on $GeI_6{}^{4-}$ corner-sharing octahedra. In contrast, larger A-site cations (e.g., guanidinium, trimethylammonium) result in the formation of one-dimensional chain-like hexagonal perovskite structures ($CsCdBr_3$-type) comprised of $GeI_6{}^{4-}$ face-sharing octahedra, causing distortions of the crystal structure. The use of the n-butylammonium ion as an A-site cation enables the formation of an orthorhombic perovskite $(C_4H_9NH_3)_2GeI_4$, which has a two-dimensional structure with perovskite sheets made of corner-sharing GeI_6 octahedra (Fig. 4.6c–f) (Stoumpos et al. 2015).

In a patent, Huang et al. claimed a PCE that is about 3% for a meso-structured perovskite solar cell design, but the experimental data are limited. This efficiency value is significantly lower than the theoretically feasible PCE value of 27.9%, as estimated by Qian et al. Therefore, additional efforts are required to enhance the effectiveness of germanium-based perovskites in achieving comparative values (Huang et al. 2014; Qian et al. 2016). Despite this, germanium halide perovskites are semiconductors that can be processed in low-temperature solutions and potentially be used in photovoltaic applications. However, the potential of this material has yet to be used to its fullest extent.

4.3.2 Alkaline-Earth Metals

Alkaline-earth metal perovskites comprised of barium, magnesium, calcium, and strontium are prospective alternatives for lead because of their similar stable +2 oxidation states, abundant presence in the Earth's crust, and non-hazardous. These metals have suitable ionic radii to form perovskite structures, and inorganic A-site cations (such as K^+ and Cs^+) were extensively studied for their photoluminescence properties when doped with rare earth metal cations like Eu^{2+}, Yb^{2+}, or Tm^{2+} in alkaline-earth metal halide perovskites that has general formula ABX_3 (B = Mg, Ca, Sr; X = Cl, Br, I) (Suta et al. 2014, 2016; Dynamics et al. 2006). However, there hasn't been much research on employing alkaline-earth metal-based halide perovskites for photovoltaic applications because of their high estimated bandgaps and susceptibility to humidity.

4.3.2.1 Magnesium (Mg^{2+})

According to simulations conducted by Filip et al. and Choudhary et al., magnesium can be substituted for lead in the structure of perovskite to produce magnesium halide perovskites with direct bandgaps that can be adjusted to fall within the visible region of the electromagnetic spectrum (Filip et al. 2016; Choudhary et al. 2015) This is made possible by the low effective masses and reasonable absorption coefficients of the magnesium halide perovskites, which depend on the size of the A-site cation. The simulations revealed that the bandgap of $AMgI_3$ perovskites can be adjusted by utilizing various A-site cations, having bandgaps of 0.9 eV ($CsMgI_3$: Fig. 4.7c), 1.5 eV ($MAMgI_3$: Fig. 4.7b), and 1.7 eV ($FAMgI_3$: Fig. 4.7a) (Filip et al. 2016). Suta et al. have successfully fabricated Eu^{2+}-doped $CsMgI_3$ that exhibits a distorted perovskite structure similar to $CsNiCl_3$. The structure consists of face-sharing $MgI_6{}^{4-}$ octahedra that form linear chains along the *c*-axis and 12-fold coordinated Cs^+ ions in the anti-cuboctahedral positions (Suta et al. 2015). Despite the promising properties of magnesium halide perovskites as absorber materials in solar cells, their sensitivity toward humidity might be the reason why they have not been implemented in this application so far.

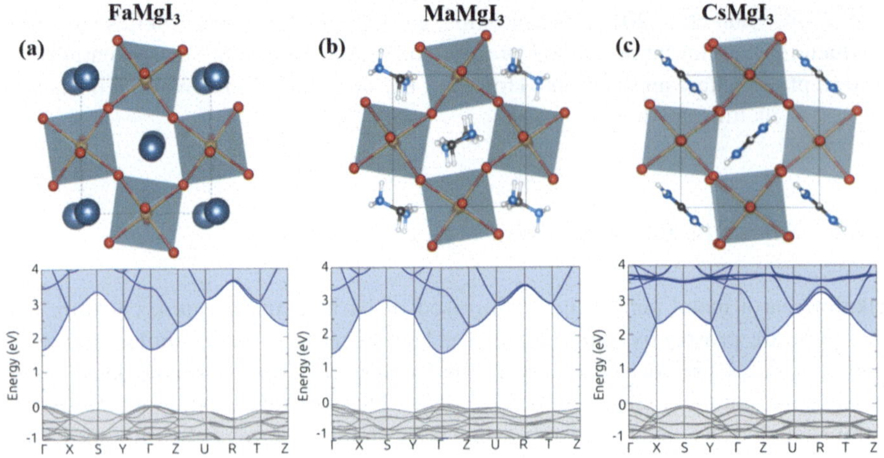

Fig. 4.7 **a–c** Typical crystal structure of Mg-based perovskites and their respective bandgap simulation using DFT, Reproduced from Ref. Filip et al. (2016), Copyright (2016) with permission of the American Chemical Society

4.3.2.2 Calcium (Ca^{2+})

Calcium is a readily available, cost-effective, and non-toxic alkaline-earth metal that can substitute lead in the perovskite structure because of its appropriate ionic radius. According to theoretical calculations by Pazoki et al., $MACaI_3$ can form a stable perovskite structure with a significantly higher calculated bandgap of 2.95 eV compared to the lead-based analogous (1.66 eV) (Pazoki et al. 2016). Uribe et al. synthesized $MACaI_3$ and $MACaI_{3-x}Cl_x$ with a pseudocubic structure using a low-temperature solution-based method from MAI combined with CaI_2 or $CaCl_2$ precursors. They determined that $MACaI_3$'s optical bandgap was 3.78 eV, which was reasonably close to the calculated bandgap of 3.4 eV (Uribe et al. 2016). Although calcium halide perovskites have a high bandgap, low mobility, and are unstable in humid environments, they could be potential candidates for charge-selective contacts but not the best choice for photovoltaic applications.

4.3.2.3 Strontium (Sr^{2+})

Strontium, an alkaline-earth metal, has a similar ionic radius to lead, making it a promising substitute for lead in perovskite structures without altering the crystal structure (Shannon et al. 1976). Currently, research is focused on $MASrI_3$ for optoelectronic applications. DFT simulations predict the production of a stable $MASrI_3$ perovskite material despite the electronegativity difference between lead and strontium. Due to the lower electronegativity of strontium and the absence of d-orbitals in the valence of Sr^{2+}, the larger bandgap of 3.6 eV for $MASrI_3$ in comparison to

Table 4.2 Typical structural and optical properties of various rare alkaline-earth-based metal halides

Crystal system	Absorber material	Bandgap (eV)	References
Tetragonal	$MABaI_3$	3.3	Pazoki et al. (2016)
	$MASrI_3$	3.6	
	$MAMgI_3$	1.5	Filip et al. (2016)
	$MACaI_3$	2.95	Uribe et al. (2016)
Orthorhombic	$CsMgI_3$	1.7	Filip et al. (2016)
	$MACaI_3$	3.78	Uribe et al. (2016)
Trigonal	$FAMgI_3$	0.9	Filip et al. (2016)

the lead analogous makes it unsuitable for use as an absorber material in solar cells. Strontium perovskites exhibit more prominent ionic interactions between the metal-halogen bond due to the greater difference in electronegativity between the metal and halogen. $MASrI_3$ can be produced through different techniques like solution-based processing, vapor phase, or layer-by-layer deposition. However, because of its hygroscopic nature, it displays poor stability in ambient conditions. Nevertheless, there is potential for $MASrI_3$ as a charge-selective contact material (Jacobsson et al. 2015; Allen et al. 1989).

4.3.2.4 Barium (Ba^{2+})

Ba^{2+} is a stable metal cation with a relatively greater ionic radius than Pb^{2+}. Applying Goldschmidt replacement rules suggests that $MABaI_3$ has a crystal structure similar to $MAPbI_3$, with a similar tolerance factor. According to DFT simulations, $MABaI_3$ is expected to form more stable perovskite materials than $MAPbI_3$, with a predicted bandgap of 3.3 eV. The high bandgap is attributed to barium's poor electronegativity and low work function. Low-temperature solution-based, or vapor-based techniques can be used to fabricate barium halide perovskites. Hence, the extreme sensitivity of barium halide perovskites toward humidity hampers their synthesis, characterization, and potential applicability in photovoltaics (Shannon et al. 1976; Kumar et al. 2016; Michaelson et al. 1977). Table 4.2 compiles the structural and optical information of perovskite materials containing alkaline-earth metal halides.

4.3.3 Transition Metals

Transition metals possess numerous oxidation states, which can cause stability issues comparable to those observed in tin and germanium perovskites. These metals have a smaller cationic radius than lead and are abundantly available. Hence, smaller monovalent A-site cations are required to form three-dimensional perovskites with transition metals following the Goldschmidt rules (Boix et al. 2015). Otherwise, two-dimensional layered perovskites that possess directional charge transport and a smaller cross-section for light absorption are produced. Additionally, certain transition metals like Cd and Hg are more toxic than lead. Despite ongoing research on transition metal halide perovskites, they exhibit less promising photovoltaic properties than lead halide perovskites. Transition metals have several oxidation states, making them structurally unstable and challenging to use as the A-site cation in perovskites. They have a smaller ionic radius compared to lead, which makes them ideal for forming three-dimensional perovskites. However, the small ionic radii of certain transition metal cations, like Cu^{2+}, Fe^{2+}, and Pd^{2+}, make it difficult for them to form three-dimensional structures. As a result, these metals tend to form lower-dimensional layered structures similar to Ruddlesden-Popper perovskites like K_2NiF_4 (Boix et al. 2015). For instance, $MA_2CuCl_xBr_{4-x}$, (Cortecchia et al. 2016) MA_2FeCl_4, (Mostafa et al. 1971) and MA_2PdCl_4 (Huang et al. 2016) are examples of such lower dimensional perovskites (Fig. 4.8a–d).

Several transition metals, including nickel, cobalt, copper, palladium, zinc, and mercury, have been proposed as suitable alternatives to lead in perovskite structures. Various lead-free perovskite materials based on transition metal halides have been reported in the literature. For example, a hydrothermal technique was used to create $CsNiX_3$ (X = Cl, Br, I) perovskites, which comprised of face-sharing NiX_6 octahedra separated by CsX_{12} cuboctahedra, from a nickel-based perovskite with a $BaNiO_3$ structure. Low electronic bandgaps and dispersive band edges make $CsNiCl_3$ and $CsNiBr_3$ promising for photovoltaic applications. Perovskites made of cobalt and iron can also be created using the hydrothermal synthesis technique. In addition, layered perovskite structures of bis-(alkylammonium) metal(II) tetrahalide and (α,ω-)polymethylene diammonium metal(II) tetrahalide were investigated (Arend et al. 1978). Lead-based materials have the same toxicity issues as mercury and cadmium halide perovskites. However, an unusual photovoltaic effect and ferroelectric behavior in a hybrid cadmium halide perovskite has recently been observed. Despite the hazardous problem of cadmium-based compounds, this material has the potential for photovoltaic applications, as evidenced by the high V_{OC} (32 V) of a bulky crystal (Ye et al. 2014).

Chalcogenide perovskites, a new group of ABX_3 perovskite structures, have recently attracted attention for their potential in solar cells. In these structures, alkaline-earth metals occupy the A site, divalent transition metals occupy the B site, and S^{2-} or Se^{2-} occupy the X site. According to DFT calculations, some of these compounds, such as $CaZrSe_3$ and $BaZrS_3$, have narrow bandgaps and are promising for use in solar cell devices. Replacing the alkaline-earth metals with MA^+ in chalcogenide perovskites has been proposed as a strategy to improve their stability (Tiwari et al. 2021). The density of states at the VB and CB edges is majorly contributed by S/Se orbitals and divalent transition metal orbitals, respectively, in these compounds. Transition metal and europium halide perovskites' structural and optical characteristics and their potential to function as absorber elements in solar cells are summarized in Table 4.3.

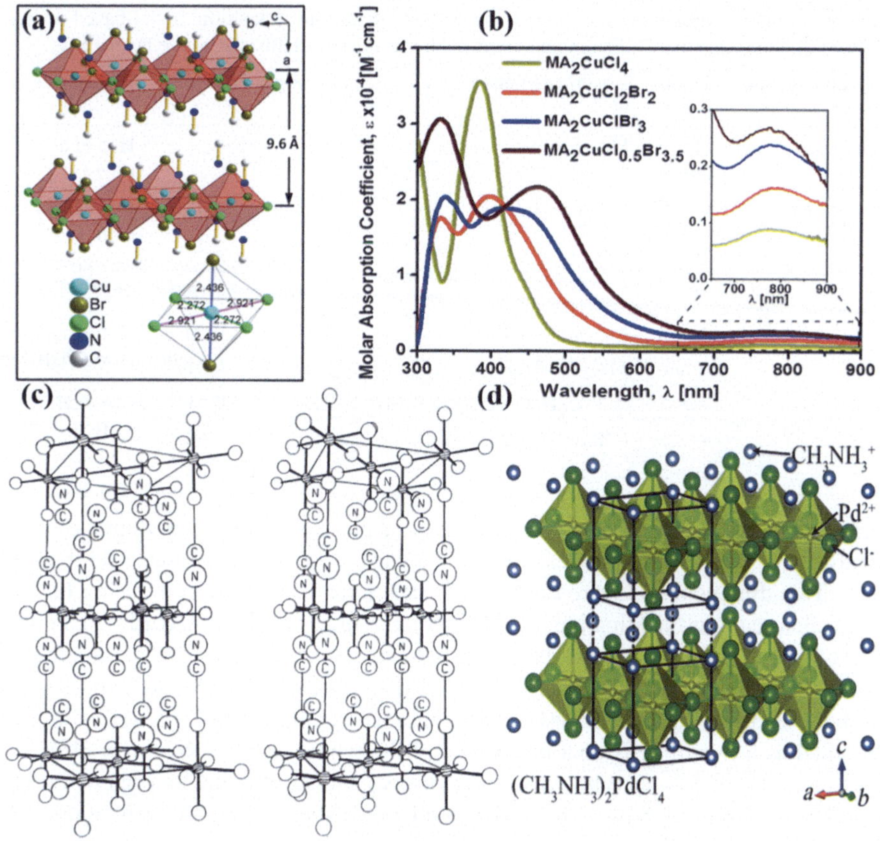

Fig. 4.8 **a, b** Typical Ruddlesden-Popper perovskite-like layered structure of $MA_2CuCl_xBr_{4-x}$ and their absorption spectrum, Reproduced from Ref. Cortecchia et al. (2016), Copyright (2016) with permission of the American Chemical Society. **c** MA_2FeCl_4, Reproduced from Ref. Mostafa et al. (1971), Copyright 1971) with permission of the American Physical Society. **d** MA_2PdCl_4 layered crystal structure, Reproduced from Ref. Huang et al. (2016), Copyright (2016) with permission of WILEY-VCH Verlag GmbH & Co. KGaA, Weinheim

4.3.4 Lanthanides/Actinides

According to Goldschmidt rules, Eu^{2+}, Dy^{2+}, Yb^{2+}, Tm^{2+}, and Np^{2+} cations of lanthanide and actinide can replace Pb^{2+} in the tetragonal perovskite lattice. Lanthanides are abundant and non-hazardous, making them a potential replacement for lead (Volonakis et al. 2016). Lanthanide compounds also have similar vapor temperatures to the lead system, making them viable options for vapor deposition. In lanthanide iodide perovskites, the VB and CB edges are made up of

Table 4.3 Typical structural and optical properties of various transition metal-based halide perovskites and respective PCEs obtained with such transition metal-based active layer PSCs

Crystal structure	Space group	Absorber material	Bandgap	PCE	References
Tetragonal	I4/mmm	MA_2FeCl_4	–	–	Han et al. (2014)
Monoclinic	$P12_1/a1$	MA_2CuCl_4	2.48	–	Cortecchia et al. (2016)
Orthorhombic	Pccn	MA_2FeCl_4	–	–	Mostafa (1977)
	Acam	$MA_2CuCl_{0.5}Br_{3.5}$	1.80	0.0017	Cortecchia et al. (2016)
	Acam	$MA_2CuCl_2Br_2$	2.20	0.017	
–	–	$(p\text{-}F\text{-}C_6H_5C_2H_4NH_3)_2CuBr_4$	1.74	0.51	Cui et al. (2015)
–	–	$(CH_3(CH_2)_3NH_3)_2CuBr_4$	1.76	0.63	

lanthanide *f*-electrons and, correspondingly, *d*-electrons since the number of *f*-electrons in various lanthanides does not contribute to the outermost shell electrons. Two-dimensional layered perovskite structures can be made by using bigger monovalent cations. Lanthanide-based perovskite materials like $MAEuI_3$, $MAYbI_3$, $MADyI_3$, $MATmI_3$, and $(C_4H_9NH_3)_2EuI_4$ have been investigated experimentally and theoretically (Mitzi and Liang 1997) Lanthanides with a 3+ oxidation state can be used for a double perovskite structure. The compound $(C_4H_9NH_3)_2EuI_4$ is classified as a two-dimensional perovskite of the type A_2BX_4 with a layered structure composed of BX_6 octahedra sharing corners and surrounded by organic butylammonium cations on both sides of the metal halide sheets. At a low temperature of approximately 140–160 °C, this material had been synthesized by a solid-state reaction between $C_4H_9NH_2{\cdot}HI$ and EuI_2. Due to the strong affinity of Eu^{2+} ions to bind solvent molecules and their instability toward oxidation, solution-based synthesis methods are not ideal for this compound. Nonetheless, both perovskite structures contain Eu^{2+} ions coordinated by six iodine atoms arranged in an octahedral shape (Morss et al. 1970).

In ABX_3-type perovskites, rare earth metal ions are frequently employed as dopants. $CsBX_3$ (B = Mg, Ca, Sr; X = Cl, Br, I) are alkaline-earth metal halide perovskites that have been investigated for their optical properties through doping with rare earth metal ions like Eu^{2+}, Tm^{2+}, and Yb^{2+}. $CsBI_3:Eu^{2+}$ and $CsBBr_3:Eu^{2+}$ (B = Mg, Ca, and Sr) are examples of host compounds where divalent Eu^{2+} metal cations take up the sixfold, octahedrally coordinated alkaline-earth metal sites (Suta et al. 2014). Quaternary halide double perovskites have also used all lanthanides and actinides. However, research on their photovoltaic properties has not yet been published as of the date of writing this book.

4.4 Mono-, Tri-, and Tetra-valent Substitutions

Another strategy for developing lead-free perovskite materials is hetero-valent substitution, which involves substituting the divalent lead cation using a cation of a distinct valence state, such as mono-, tri-, or tetravalent cations. However, substituting hetero-valent cations is a complex process due to the differences in valence states.

According to Goldschmidt's rules, monovalent metals without charge compensation cannot replace lead in the perovskite structure. However, they can be partially added into the lattice as dopants in lattice or interstitial sites to modify the material properties. The enhancement of charge mobility and conduction is the driving force behind this inclusion. Such additions can change the distortion of the nearest PbI_6 octahedron, shift the bands, and affect the defect density. Abdi-Jalebi et al. added monovalent cation additives (Cu^+, Ag^+, and Na^+) to the $MAPbI_3$ precursor solution to balance the hole and electron conductivities in the material (Abdi-Jalebi et al. 2016). AgI and NaI additions showed increased voltage and current, which the researchers ascribed to improved crystal formation and bulk charge transport. Recently, a cation cascade mixed perovskite has been reported to have enhanced stability and performance due to the implementation of additional monovalent cation precursors such as CsI, KI, and RbI (Saliba et al. 2016). These precursors can replace the A-site: MA^+ or FA^+ cations, which can also act as interstitial defect modifiers in the perovskite lattice. In order to change the characteristics of the material, divalent and trivalent dopant additives are also being used. However, it is sometimes obvious whether they accumulate at interfaces or in the bulk of the perovskite. Additionally, it has been demonstrated that a mixture of monovalent and trivalent cations can replace lead in a double perovskite lattice.

In halide perovskite solar cell applications, trivalent metal cations were investigated as potential lead replacements, with Bi^{3+} being the most extensively studied. Due to its oxidation state and preferred chemical bonds, Bi^{3+} cannot fit within the tetragonal structure of perovskites. Instead, the presence of Bi^{3+} leads to the hexagonal structure $A_3Bi_2I_9$ (where A can be Cs or MA), with the BiI_6 octahedra sharing faces and lattice constants a = 8.39 Å and c = 20.98 Å (Kamminga et al. 2017). Iodine states make up the valence band edge states, whereas hybridized bismuth-iodine states and a second conduction band with a gap of 500 meV from the first, which is primarily made of Cs states, make up the lowest states in the conduction band. Bismuth perovskite materials can be made utilizing vapor deposition and solution processing techniques, and they have high-born effective charges. At 160 K, there is a phase change to a monoclinic crystal structure with the space group C2/c.

The unit cell's chemical bonding is such that the charge density planes within the ab-plane exhibit good continuity. At the same time, there is no bonding between the Bi-I networks in the c-axis direction, leading to a layered perovskite with directional charge transport. According to Kamminga and the group, this layered structure possesses a bandgap of 2.2 eV and is believed to be less toxic and more stable than 3D lead perovskite (Kamminga et al. 2017). The material is better suited for tandem applications due to the relatively large bandgap value, which restricts the maximum achievable PCE from single bandgap devices. In further investigation, the band dispersion exhibits directional transport, with the flat bands corresponding to high intrinsic effective masses in the out-of-plane directions (Hoye et al. 2016). The electronic structure of bismuth compounds is significantly affected by SOC effects, although to a lesser extent than in lead compounds. Moreover, compared to $MAPbI_3$, the directional charge transport is limited, and the dielectric constant is lower, while the exciton binding energy and absorption coefficient are higher. Consequently, the highest reported efficiency for bismuth-based PSCs is only about 1%. Experimental and theoretical values of 300–400 meV have been reported, and ab initio calculations have suggested that the high exciton binding energy arises from the low dielectric constant. However, these properties may be advantageous for light-emitting applications such as in $MA_3Bi_2Br_9$. Excess iodide was added to the precursors to improve the material's properties; this led to the creation of the new material $Cs_3Bi_2I_{10}$, which possesses a bandgap of 1.77 and a substantially greater absorption coefficient than $Cs_3Bi_2I_9$ (Kawai et al. 1996). $AgBiI_4$ and Ag_3BiI_6 have undergone additional modifications that have increased light absorption and efficiencies, with the maximum efficiency being 4.3% (Turkevych et al. 2017).

4.4.1 *Thallium (Tl^+ or Tl^{3+})*

Thallium, a p-block metal, has a Tl^+ cation that shares the same electronic configuration with Pb^{2+} ($6s^2\ 6p^0$). However, due to the charge neutrality issue between cationic and anionic species, substituting the divalent Pb^{2+} metal cation with the monovalent Tl^+ cation is impossible in ABX_3-type perovskites. Nonetheless, thallium can be incorporated into the perovskite structure using the mixed-valence approach involving Tl^+ ($6s^2$) and Tl^{3+} ($6s^0$). As seen in $CsTl^+_{0.5}Tl^{3+}_{0.5}X_3$ (X = F, Cl), the mono- and trivalent thallium cations can coexist in a charge-ordered perovskite structure (Fig. 4.9a) (Retuerto et al. 2013). Retuerto et al. and Yin et al. investigated the superconductive behavior of this type of thallium halide perovskites. Experimental observations on the optical properties of $CsTlCl_3$ revealed an optical bandgap of approximately 2.5 eV, as shown in Fig. 4.9b–d (Kamminga et al. 2017; Yin et al. 2013).

A mixed thallium-bismuth halide perovskite, $MATl_{0.5}Bi_{0.5}I_3$, has been developed as a lead-free hybrid perovskite alternative to $MAPbI_3$ by substituting Pb^{2+} metal cation units with Tl^+/Bi^{3+} hetero-valent ionic pairs. Using DFT analysis, Giorgi et al. studied the structural and electronic characteristics of this perovskite, which resulted in a direct bandgap of 1.68 eV. Despite the potential of $MATl_{0.5}Bi_{0.5}I_3$ as an alternative solar cell material, its toxicity limits its application in photovoltaic devices (Giorgi and Yamashita 2015). Although this compound has promising optical properties, as summarized in Table 4.4, thallium-based compounds are not considered a viable substitute for lead-based perovskites in photovoltaic applications.

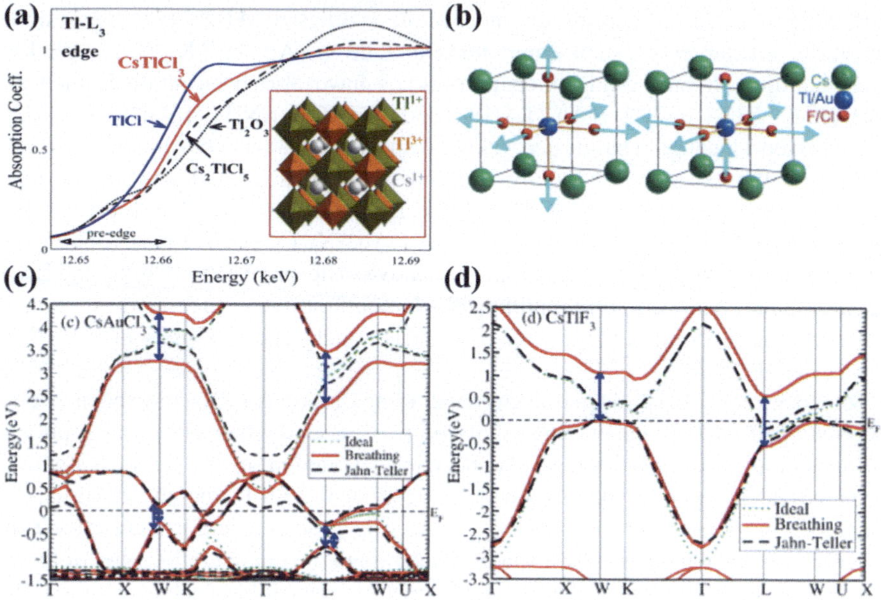

Fig. 4.9 **a** Absorption spectrum of $CsTl^{+}_{0.5}Tl^{3+}_{0.5}Cl_3$, Reproduced from Ref. Retuerto et al. (2013), Copyright (2013) with permission of the American Chemical Society. **b–d** Typical crystal structure and bandgap estimation of $CsAuCl_3$ and $CsTlF_3$ perovskites, respectively, Reproduced from Ref. Yin et al. (2013), Copyright (2013) with permission of IOPScience

Table 4.4 Typical structural and optical properties of various Tl-based halide perovskites

Crystal system	Space group	Absorber material	Bandgap (eV)	References
Triclinic	P $\bar{1}$	$[NH_3(CH_2)_7NH_3]_2[(Au^+I_2)(Au^{3+}I_4)(I_3)_2]$	0.95	Castro-Castro and Guloy (2003)
Monoclinic	C2/m	$[NH_3(CH_2)_8NH_3]_2[(Au^+I_2)(Au^{3+}I_4)(I_3)_2]$	1.14	
Tetragonal	I4/m	$(CsTl^{+}_{0.5}Tl^{3+}_{0.5}F_3Cl_3)$	2.5	Yin et al. (2013)
	I4/mmm	$Cs_2Au^IAu^{III}I_6$	1.31	Riggs et al. (2012), Liu et al. (1999)
	I4/mmm	$Cs_2Au^IAu^{III}Cl_6$	2.04	
	I4/mmm	$Cs_2Au^IAu^{III}Br_6$	1.60	

4.4.2 *Gold (Au^+ or Au^{3+})*

Gold halide perovskites can be synthesized using a mixed-valence method similar to thallium-based analogous. Monovalent Au_+ ($5d^{10}$, t^62g e^4g) and trivalent Au^{3+}

($5d^8$, t^62g e^2g) metal cations are needed to synthesize ABX_3-type perovskite structures. Examples of such compounds include $Cs_2Au^+Au^{3+}X_6$ (X = Cl, Br, I). In addition, hybrid gold halide perovskites have also been studied, such as $[NH_3(CH_2)_7NH_3]_2[(Au^+I_2)(Au^{3+}I_4)(I_3)_2]$ and $[NH_3(CH_2)_8NH_3]_2[(Au^+I_2)(Au^{3+}I_4)(I_3)_2]$. Mixed-valent gold halide perovskites display two distinct coordination spheres because of the existence of monovalent and trivalent metal cations. Specifically, the coordination of Au^+ is twofold and linear, while that of Au^{3+} is fourfold and square-planar. The crystal structure of $Cs_2Au_+Au^{3+}X_6$ (X = Cl, Br, I) is obtained from a distorted ABX_3-type perovskite, where BX_2 (linear $[Au^+X_2]^-$ unit) and BX_4 (square-planar $[Au^{3+}X_4]^-$ unit) building blocks combine to form a three-dimensional perovskite network with $Cs_2[Au^+X_2][Au^{3+}X_4]$ stoichiometry (Riggs et al. 2012; Saparov et al. 2016).

Mixed-valent gold halide perovskites, including $Cs_2Au^+Au^{3+}X_6$, have been extensively studied for their superconductivity properties, while their electronic, optical, and structural characteristics have also been investigated in several research studies. In mixed-valent systems like $Cs_2Au^+Au^{3+}X_6$ (X = Cl, Br, I), the choice of halide counterion is crucial for the bandgap engineering process since it can be used to lower the optical bandgap. When Liu et al. examined the optical reflectivity of these perovskites, they discovered that the corresponding optical bandgaps were 2.04 eV (X = Cl), 1.60 eV (X = Br), and 1.31 eV (X = I) (Riggs et al. 2012). Of note, $Cs_2Au^+Au3^+X_6$ is a very attractive absorber material for photovoltaic applications due to its nearly optimal bandgap, which approaches the Shockley-Queisser limit. Additionally, $Cs_2Au^+Au^{3+}X_6$'s potential as an alternative for lead-based equivalents is increased by the fact that its three-dimensionally deformed ABX_3-type perovskite structure is similar to that of lead-based counterparts.

The optical characteristics of layered hybrid gold halide perovskites that are two-dimensional like $[NH_3(CH_2)_7NH_3]_2[(Au^+I_2)(Au^{3+}I_4)(I_3)_2]$ and $[NH_3(CH_2)_8NH_3]_2[(Au^+I_2)(Au^{3+}I_4)(I_3)_2]$ were investigated by Castro-Castro et al. (2003) The researchers determined that the bandgaps of these materials were 0.95 and 1.14 eV, which is less than the bandgap of the three-dimensional $Cs_2Au^+Au^{3+}I_6$ perovskite (1.31 eV). This is an unusual trend since lower dimensional perovskites often display larger bandgaps. However, the additional electronic interactions between the $[Au^+I_2]^-$ and $[Au^{3+}I_4]^-$ units and I_3^- ions that are not present in $Cs_2Au^+Au^{3+}I_6$ could explain these unexpectedly low bandgaps.

4.4.3 Antimony (Sb^{3+})

Antimony halide perovskites have grown to be a promising replacement for lead-based perovskite semiconductors in photovoltaic applications owing to their potential to overcome toxicity and chemical stability concerns. The trivalent Sb^{3+} metal cation is isoelectronic to Sn^{2+} and has an identical s^2 valence electronic configuration as Pb^{2+} but possesses a smaller ionic radius than Pb^{2+} and Sn^{2+} metal cations. Additionally, Sb^{3+} has a comparable electronegativity to Sn and Pb. These characteristics make antimony halide perovskites an attractive candidate for optoelectronic applications. The selection of cationic and anionic species has a major impact on the dimensionality and structural chemistry of antimony halide perovskites. Based on their crystal structures and dimensionality, these perovskites, which contain $Sb_2X_9^{3-}$ enneahalide ions within the anionic sub-lattice, can be categorized into three groups, as illustrated in Fig. 4.10 (Zaleski et al. 1996; Yamada et al. 1997). The selection of cationic and anionic species substantially impacts the structural chemistry and dimensionality of antimony halide perovskites. Based on the dimensionality of the crystal structures of antimony-based perovskites that contain $Sb_2X_9^{3-}$ enneahalide ions in the anionic sublattice can be classified into three types: zero-dimensional, which consist of isolated double octahedral structures arranged in dimer units; infinite one-dimensional, which are composed of double chains of corner-sharing SbX_6 octahedra; and two-dimensional, which are corner-connected SbX_6 octahedral corrugated double-layered polyanionic structures.

Hebig et al. recently conducted a study on the optoelectronic properties of $MA_3Sb_2I_9$. This dimer structure of this compound consists of discrete bi-octahedral metal halide units $Sb_2I_9^{3-}$, which share faces with BI_6 octahedra surrounded by MA^+ cations to maintain charge neutrality (Hebig et al. 2016). The complex anionic clusters are joined by N–H···I type hydrogen bonding interactions. $MA_3Sb_2I_9$ had been synthesized from MAI and SbI_3 using a solution-based deposition technique at lower temperatures (100–120 °C). The compound has a peak absorption coefficient similar to the lead-based equivalent of about 10^5 cm^{-1}. Considering a direct band transition, the optical bandgap had been found to be 2.14 eV. Peresh et al. studied the optical properties of inorganic antimony halide perovskites with an $A_3Sb_2Br_9$-type structure. They found that the bandgaps were 2.48 eV (A = Rb^+)

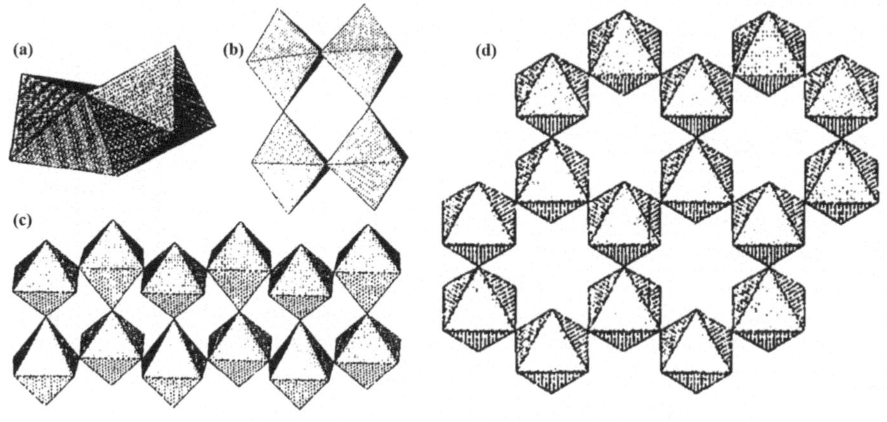

Fig. 4.10 Typical schematics of Sb-based perovskites in polyhedral representation. **a** 0-D dimer phase of face sharing octahedra, **b** isolated unit consisting of 4-octahedra, **c** 1-D layered chains of corner-sharing octahedra, **d** 2-D double-layered structure of corner-sharing octahedra, Reproduced from Ref. Zaleski et al. (1996), Copyright (1996) with permission of the International Union of Crystallography

and 2.30 eV (A = Cs^+), respectively. By replacing Br^- with the heavier I^-, the bandgap of $Cs_3Sb_2I_9$ can be lowered to 1.89 eV, making it a promising candidate for deployment in photovoltaic applications (Peresh et al. 2011). Saparov et al. have researched the use of $Cs_3Sb_2I_9$'s layered modification as a light absorber in PSCs. The cells were designed with a device architecture of FTO/*c*-TiO_2/$Cs_3Sb_2I_9$/PTAA/Au(PTAA:poly[bis(4-phenyl)(2,4,6-trimethylphenyl)amine]).

Unfortunately, the material demonstrated suboptimal photovoltaic performance, characterized by a maximum open-circuit voltage (V_{OC}) of 300 mV, a short-circuit current density (J_{SC}) of less than 0.1 mA cm^{-2}, and an overall PCE of less than 1%. In $A_3Sb_2X_9$-type perovskites, the bandgap slightly shifts toward shorter wavelengths when Cs^+ is switched out for Rb^+ (Saparov et al. 2015). Compared to a bandgap value of 2.05 eV for the Cs-based analogous, experimental analysis revealed an indirect bandgap of 2.1 eV and a direct transition at 2.24 eV for $Rb_3Sb_2I_9$. Furthermore, $Rb_3Sb_2I_9$ demonstrated an absorption coefficient exceeding 1×10^5 cm^{-1}, which is equivalent to lead-based systems. Harikesh et al. conducted research on solution-processed $Rb_3Sb_2I_9$ perovskite absorbers in solar cells utilizing an FTO/*c*-TiO_2/*mp*-TiO_2/$Rb_3Sb_2I_9$/poly-TPD/Au device architecture (poly-TPD: poly[N, N′-bis(4-butylphenyl)-N, N′-bisphenylbenzidine]). The fabricated solar cells showed a V_{OC} of 0.55 V, a J_{SC} of 2.12 mA cm^{-2}, and a fill factor (FF) of 57%, leading to a PCE of 0.66%. These findings represent encouraging advancements in exploring alternative lead-free perovskite semiconductors for solar cell applications (Harikesh et al. 2016).

Volonakis et al. conducted a theoretical study on antimony halide double perovskite semiconductors represented by the fundamental formula $A_2B^+B^{3+}X_6$. These materials are structured when Pb^{2+} is swapped out for an equal amount of monovalent and trivalent cations; double perovskite structures are formed in the elpasolite structure while maintaining charge neutrality (Volonakis et al. 2016). Antimony halide perovskites' structural and optical characteristics and how well they work as absorber materials in solar cells are summarized in Table 4.5.

4.4.4 *Bismuth (Bi^{3+})*

Bi^{3+} is a promising substitute for lead and tin due to its isoelectronic nature with Pb^{2+} ions, similar electronegativity, and comparable ionic radius. However, due to the various valence states, the trivalent Bi^{3+} ion cannot simply substitute the divalent Pb^{2+} ion in the perovskite structure. Regarding connectivity and dimensionality, bismuth halide perovskites show a wide range of structural diversity, with structures ranging from zero-dimensional dimer units to three-dimensional double perovskite frameworks, such as the elpasolite structure.

$MA_3Bi_2I_9$ is the bismuth halide perovskite that has been extensively studied for optoelectronic applications. This perovskite has BiI_6 octahedra that share faces, forming $Bi_2I_9{}^{3-}$ dimers that are isolated and surrounded by MA^+ cations in a random disordered. The bi-octahedral clusters are connected through N–H···I hydrogen bonds (Dammak et al. 2015; Öz et al. 2016) When the perovskite is cooled, phase transitions

Table 4.5 Typical structural and optical properties of various Sb-based halide perovskites

Crystal system	Space group	Absorber material	Bandgap (eV)	PCE (%)	References
Cubic	Fm3m	Cs_2CuSbX_6 (Cl, Br, I)	2.1 (Cl), 1.6 (Br), 0.9 (I)	–	Volonakis et al. (2016)
		Cs_2AgSbX_6 (Cl, Br, I)	2.6 (Cl), 1.9 (Br), 1.1 (I)	–	
		Cs_2AuSbX_6 (Cl, Br, I)	1.3 (Cl), 0.7 (Br), 0 (I)	–	
Hexagonal	P63/mmc	$MA_3Sb_2I_9$	2.14	0.5	Hebig et al. (2016)
		$Cs_3Sb_2I_9$	1.89–2.4	< 1	Saparov et al. (2015)
Trigonal	P3m1	$Rb_3Sb_2Br_9$	2.48	–	Peresh et al. (2011)
		$Cs_3Sb_2Br_9$	2.30	–	
		$Cs_3Sb_2I_9$	2.05	< 1	Saparov et al. (2015)
Monoclinic	Pc	$Rb_3Sb_2I_9$	2.1	0.66	Harikesh et al. (2016)

take place from a hexagonal crystal structure (space group: P63/mmc) at 300 K to a monoclinic crystal structure (space group: C2/c) at 160 K and another first-order phase transition (monoclinic, space group: P21) occurs at 143 K (Kamminga et al. 2017). These phase transitions are brought on by dipolar ordering of the organic cation and in-plane ordering of the metal's lone pair. Additionally, $MA_3Bi_2I_9$ has intense absorption at around 500 nm, a pre-edge absorption peak at 2.51 eV that indicates the presence of intrinsic excitons, and a high optical absorption coefficient in the order of 105 cm^{-1}, which is equivalent to that of lead-based perovskites. According to a study by Zhang et al., the efficiency of PSCs can be improved by employing a modified annealing technique of the ETL and ITO as the transparent interface in place of FTO. Furthermore, the processing conditions and the ETL's structure, which can be either planar or meso-structured, have a substantial impact on the morphology of the active layer and, as a result, the performance of the photovoltaic system. For instance, the PCE of the solar cell was enhanced from 0.14% to 0.42% by changing the architecture from planar (ITO/*c*-TiO_2/$MA_3Bi_2I_9$/Spiro-MeOTAD/MoO_3/Ag) to meso-structured (ITO/*c*-TiO_2/*mp*-TiO_2/$MA_3Bi_2I_9$/Spiro-MeOTAD/MoO_3/Ag) (Zhang et al. 2016b).

Examples of two-dimensional layered defect perovskites synthesized by solution-based or solid-state reactions are $K_3Bi_2I_9$ and $Rb_3Bi_2I_9$. Studies have demonstrated that these compounds are more stable under immediate surrounding conditions than lead- and tin-based analogous. Electronic band structure calculations indicate that both compounds possess a direct band character and an optical bandgap of 2.1 eV. In contrast, the bigger A-site cation Cs^+ prohibits the $Cs_3Bi_2I_9$ analogous from adopting a two-dimensional perovskite structure, resulting in a zero-dimensional perovskite structure that possesses distinct optoelectronic properties, as previously explained. Furthermore, $CsBi_3I_{10}$ exhibits superior film-forming characteristics to $Cs_3Bi_2I_9$, resulting in more uniform and smoother layers without any pinholes, making it a favorable option for photovoltaic applications. Meso-structured solar cells (FTO/*c*-TiO_2/*mp*-TiO_2/perovskite/P3HT/Ag), which demonstrated a power conversion efficiency of 0.40%, were made using $CsBi_3I_{10}$ as the absorber material. This efficiency is considerably greater than that of the solar cells using $Cs_3Bi_2I_9$ (0.02%) and BiI_3 (0.07%) under the same device architecture. However, the PCE is lower than the $Cs_3Bi_2I_9$-based solar cells reported by Park et al., which has shown a PCE of 1.09% (Park et al. 2015).

Bismuth-containing three-dimensional perovskite structures were achieved in quaternary double perovskites that have a basic formula unit of $A_2B^+B^{3+}X_6$, where Pb^{2+} is replaced by a mixture of monovalent Bi^+ and trivalent Bi^{3+} cations through heterovalent substitution. Elpasolite, a type of double perovskite structure, consists of corner-sharing B^+X_6 and $B^{3+}X_6$ octahedra arranged in a rock-salt ordered cubic structure along the three crystallographic axes that form a three-dimensional network with mono- and trivalent metal ions that occupy the B^+ and B^{3+} sites, accordingly. A-site cations like Cs^+ or MA^+ are present in the cuboctahedral voids of the elpasolite structure (Slavney et al. 2016). The synthesis, crystal structure, and optical and electrical properties of different bismuth-based halide double perovskites have all been studied theoretically and experimentally. For instance, Cs_2AgBiX_6 (X = Cl, Br) perovskites can be prepared using solid-state or solution-based methods, have a crystalline structure that resembles elpasolite, and demonstrate enhanced stability against moisture and heat as compared to lead-based halide perovskites. Nevertheless, even after several weeks of exposure to air and light, $Cs_2AgBiBr_6$ still suffers degradation. Both indirect semiconductors like $Cs_2AgBiCl_6$ and $Cs_2AgBiBr_6$, have experimental bandgaps of 1.95–2.19 eV for $Cs_2AgBiBr_6$ and 2.2–2.77 eV for $Cs_2AgBiCl_6$, respectively. Table 4.6 summarizes the information on the crystal structure, optical characteristics, and photovoltaic performance of bismuth halide perovskites.

4.5 Mixed Chalcogenides and Halides

Although there have been advances in developing lead-free perovskite semiconductors for solar cell applications, these alternatives have yet to achieve the same efficiency level and still face chemical stability and toxicity challenges. One possible solution is to replace some or all of the halide anions with chalcogenide anions in the perovskite structure. This can be achieved using the split-anion method, which involves partially substituting halides with chalcogenides in metal halide perovskites of the ABX_3-type to form mixed chalcogenide-halide perovskites having AB(Ch, X)$_3$ as the general formula (Sun et al. 2016c). These mixed compounds are thought to be more stable in ambient conditions than purely halide-based perovskites because of the stronger covalent bonding between the metal and chalcogenide atoms.

Table 4.6 Typical structural and optical properties of various Bi-based halide perovskites

Crystal system	Space group	Absorber material	Bandgap (eV)	PCE (%)	References
Monoclinic	$P2_1/c$	$(NH_4)_3Bi_2I_9$	2.04	–	Sun et al. (2016b)
	$P2_1/n$	$K_3Bi_2I_9$	2.10	–	Lehner et al. (2015)
		$Rb_3Bi_2I_9$	1.89–2.1	–	
Hexagonal	$P6_3/mmc$	$MA_3Bi_2I_9$	1.94–2.11	0.42	Abulikemu et al. (2016)
		$MA_3Bi_2I_{9-x}Cl_x$	2.40	0.003	Park et al. (2015)
		$Cs_3Bi_2I_9$	1.8–2.2	1.09	
Trigonal	P3m1	$Cs_3Bi_2Br_9$	2.50	–	Peresh et al. (2011)
	R3m	MA_2KBiCl_6	3.04	–	Wei et al. (2017)
Orthorhombic	Pnma	$Rb_3Bi_2Br_9$	2.62	–	Peresh et al. (2011)
		$HDABiI_5$	2.05	0.027	Fabian and Ardo (2016)
	Pbca	$(C_{10}H_7NH_3)BiI_4$	2.32	–	Hrizi et al. (2012)

Halogen anions (X = Cl, Br, I) are partly replaced with chalcogenides (Ch = S, Se, Te), with one chalcogenide per formula unit in order to produce I–III–VI–VII_2-type semiconductors with the formula $MABiChX_2$. The calculated direct bandgaps of these materials are in the range of 1.24–2.00 eV (Hong et al. 2016). Among them, $MABiSeI_2$ and $MABiSI_2$, with direct bandgaps of 1.3 and 1.4 eV, accordingly, were found to be effective absorber materials. In a study by Sun et al., $CsSnS_2Cl$ was investigated as a potential light-absorbing material for photovoltaic applications, serving as an example for inorganic mixed-anion perovskites. By using hybrid functional calculations, the indirect bandgap of $CsSnS_2Cl$ in its deformed perovskite phase was estimated to be around 1 eV. Additionally, it was anticipated that its optical absorption properties would be more effective than those of $CsSnI_3$, suggesting its potential as a promising candidate for photovoltaic applications. Even though the mixed chalcogenide-halide strategy has not yet produced novel absorber materials, the split-anion method remains promising for future investigations due to the wide range of viable compositions for novel I–III–VI–VII_2, II–II–VI–VII_2, I–IV–VI_2–VII, or II–III–VI_2–VII semiconductors.

In recent decades, a significant amount of research has been dedicated to investigating crystal structures and physicochemical properties of various metal chalcogenide perovskites. In one study, Perera et al. used carbon disulfide to sulfurize oxide perovskite analogous at high temperatures in order to produce chalcogenide perovskites, such as $BaZrS_3$, $SrTiS_3$, $CaZrS_3$, and $SrZrS_3$. UV–Vis and photolumi nescence tests revealed direct bandgaps of 1.73 and 1.90 eV for $BaZrS_3$ and $CaZrS_3$, respectively, indicating their potential as absorber materials for photovoltaic applications. $BaZrS_3$, for instance, can be adjusted in composition through partial replacement of sulfide ions with oxygen ions, leading to the development of transition metal oxysulfide perovskites such as $BaZr(O_xS_{1-x})_3$, which exhibit a wide range of bandgaps ranging from 1.73 eV (like $BaZrS_3$) to 2.87 eV (Perera et al. 2016). Additionally, the investigated transition metal chalcogenide perovskite materials demonstrate enhanced chemical stability in an ambient environment in comparison to metal halide perovskite analogous because of the higher degree of covalent bonding nature of the metal–chalcogenide bond. Additionally, future photovoltaic applications may consider oxide perovskites and several materials with appropriate optical characteristics (such as $BiFeO_3$) that have previously been studied in photovoltaic systems.

Tin-based perovskites have drawn the most attention as they exhibit the highest PCE values of all the lead-free perovskite materials that have been reported, with PCEs of around 6% achieved with $MASnI_3$ and $FASnI_3$. Though tin perovskites are less stable than lead-based perovskite absorbers in terms of stability, development was accomplished in this area, and a lifespan test lasting 77 days with no efficiency reduction was reported. This demonstrates the potential of perovskites based on tin as highly promising materials for developing sustainable and low-cost lead-free solar cells. On the other hand, Germanium halide perovskites exhibit bandgaps comparable to those of lead-based compounds. Still, they are less stable chemically and have been far less studied than tin-based perovskites. This could be the cause of the PCEs of germanium perovskite-based solar cells continuing to be significantly lower.

Bismuth halide perovskites exhibit significant promise as an alternative due to their extensive structural variety ranging from zero-dimensional to three-dimensional structures, and their ability to possess tunable bandgaps in the visible range, with PCE values already surpassing 1%. Compared to germanium- or tin-based perovskites, bismuth perovskites demonstrate better ecological stability. Additionally, due to their limited chemical instability and toxicity, metal chalcogenide perovskite semiconductors offer an effective substitute for lead-based systems. Novel absorber materials can be developed with a wide range of bandgap engineering by adjusting stoichiometry and compositions, for instance, through a split-anion or anion alloying strategy to create mixed halide-chalcogenide compounds. Theoretical calculations suggest that metal chalcogenide perovskite semiconductors have significant direct bandgaps and superior optical absorption properties in the visible range compared to lead-based substitutes, underscoring their prospectiveness for photovoltaic applications.

4.6 Conclusion

Tin-based perovskites have garnered significant attention and have been the subject of extensive research among the various lead-free perovskite materials investigated thus far. They have demonstrated the maximum PCE among all alternative PSCs studied. The efficiency of solar cells utilizing double perovskites $A_2B^+B^{3+}X_6$, vacancy ordered double perovskites A_2BX_6, or 2D perovskite derivatives $A_3B_2X_9$ for solar cell applications has not exceeded 4% thus far, significantly lower than the impressive 25.8% attained using Pb-based halide perovskites. Among these alternative perovskite compositions, tin-iodides $ASnI_3$ have shown the most promise, with a current record efficiency of 14.6%. Hence, there is still room for advancement as the theoretical limit for thick $ASnI_3$ layers under AM1.5G irradiation is predicted to be as high as 32.3%. One drawback of tin-based perovskites is their reduced stability compared to lead-based counterparts, considering their exposure to the surrounding environment, mainly because of the rapid oxidation of Sn^{2+}. Nonetheless, recent progress has been made in enhancing stability, achieving over 1300 h of operational stability in a nitrogen environment through chemical engineering approaches, including a secondary ammonium salt and a powerful reducing agent. Another strategy to improve stability involves incorporating Sn/Ge mixed perovskite absorbers, although their efficiency still needs significant enhancement. It is worth noting that Sn-based iodide perovskites have the potential to serve as the foundation materials for tandem perovskite/PSCs or other applications that need low bandgaps because of their low electronic bandgap. In conclusion, while the efficiency of alternative perovskite compositions for solar cells remains relatively low, tin-iodide perovskites show promise with ongoing efforts to improve stability and efficiency, making them potential candidates for future advancements in solar cell technology.

Alkaline-earth metals like Ba, Mg, Ca, and Sr are often considered viable substitutes for lead in perovskite structures because of their similar ionic radii. Among these metals, magnesium iodide perovskites have shown promise as they possess a tunable visible bandgap ranging from 0.9 to 1.7 eV. However, perovskites based on calcium, strontium, and barium halides may not be suitable alternatives for lead halide perovskite semiconductors in photovoltaic applications. This is primarily because of their high bandgaps, which range from 2.95 to 3.6 eV, making them less efficient in converting sunlight into electricity. Additionally, these perovskite materials exhibit sensitivity to humidity, which can negatively impact their stability and performance. In summary, while magnesium iodide perovskites show potential for photovoltaic applications, further research and development are needed to overcome the challenges associated with high bandgaps and humidity sensitivity in Ca-, Sr-, and Ba-based halide perovskites.

Metal chalcogenide perovskite semiconductors offer a viable alternative to overcome the limitations of lead-based systems' toxicity and chemical instability. Through innovative design of materials and precise control of stoichiometry and compositions, novel absorber materials can be developed by employing strategies such as split-anion or anion alloying approaches for forming mixed halide-chalcogenide compounds. These approaches make it possible to engineer bandgaps over a wide range, facilitating the development of metal chalcogenide perovskite semiconductors with remarkable properties. Theoretical calculations have demonstrated that metal chalcogenide perovskite semiconductors have the potential to be used in photovoltaics. In comparison to their lead-based counterparts, these calculations not only predicted excellent direct bandgaps but also enhanced optical absorption properties within the visible region. This highlights the potential of metal chalcogenide perovskite semiconductors as highly efficient materials for solar energy conversion.

In conclusion, the limitations of lead-based systems are overcome by the exploration of metal chalcogenide perovskite semiconductors. Novel absorber materials with enhanced properties can be achieved by utilizing bandgap engineering and innovative materials design techniques. The theoretical predictions of favorable bandgaps and increased optical absorption in the visible range further emphasize the prospects of metal chalcogenide perovskite semiconductors for advancing photovoltaic technologies.

References

Abdi-Jalebi M, Dar MI, Sadhanala A et al (2016) Impact of monovalent cation halide additives on the structural and optoelectronic properties of $CH_3NH_3PbI_3$ perovskite. Adv Energy Mater 6. https://doi.org/10.1002/aenm.201502472

Abulikemu M, Ould-Chikh S, Miao X et al (2016) Optoelectronic and photovoltaic properties of the air-stable organohalide semiconductor $(CH_3NH_3)_3Bi_2I_9$. J Mater Chem A 4:12504–12515. https://doi.org/10.1039/c6ta04657f

Aftab A, Ahmad MI (2020) Role of additives SnX_2 (X = F, Cl) and anti-solvents on the microstructure of PV absorber $FASnI_3$ films. Mater Lett 275:13–15. https://doi.org/10.1016/j.matlet.2020.128071

Aftab A, Ahmad MI et al (2021) A review of stability and progress in tin halide perovskite solar cell. Sol Energy 216:26–47. https://doi.org/10.1016/j.solener.2020.12.065

Alcocer MJP, Leijtens T, Herz LM et al (2013) Electron-hole diffusion lengths exceeding trihalide perovskite absorber. Science (80-) 342:341–344. https://doi.org/10.1126/science.1243982

Allen LC et al (1989) Electronegativity is the average one-electron energy of the valence-shell electrons in ground-state free atoms. J Am Chem Soc 111:9003–9014. https://doi.org/10.1021/ja00207a003

Arend H, Huber W, Mischgofsky FH, Richter-Van Leeuwen GK (1978) Layer perovskites of the $(CnH_{2n+1}NH_3)_2MX_4$ and $NH_3(CH_2)_mNH_3MX_4$ families with M = Cd, Cu, Fe, Mn OR Pd and X = Cl or Br: Importance, solubilities and simple growth techniques. J Cryst Growth 43:213–223. https://doi.org/10.1016/0022-0248(78)90170-7

Ayaydah W, Raddad E, Hawash Z (2023) Sn-based perovskite solar cells towards high stability and performance. Micromachines 14. https://doi.org/10.3390/mi14040806

Bartel CJ, Sutton C, Goldsmith BR et al (2019) New tolerance factor to predict the stability of perovskite oxides and halides. Sci Adv 5:1–9. https://doi.org/10.1126/sciadv.aav0693

Boix PP, Agarwala S, Koh TM et al (2015) Perovskite solar cells: beyond methylammonium lead iodide. J Phys Chem Lett 6:898–907. https://doi.org/10.1021/jz502547f

Cai Q, Li H, Jiang Y et al (2018) High-efficiency perovskite solar cells based on $MAI(PbI_2)$, 1–x$(FeCl_2)$X absorber layers. Sol Energy 159:786–793. https://doi.org/10.1016/j.solener.2017.11.063

Castro-Castro et al (2003) Organic-based layered perovskites of mixed-valent Gold(i)/Gold(iii) iodides. Angew Chemie 2797–2909. https://doi.org/10.1002/ange.200350929

Castro-Castro LM, Guloy AM (2003) Organic-based layered perovskites of mixed-valent Gold(I)/Gold(III) iodides. Angew Chemie 115:2877–2880. https://doi.org/10.1002/ange.200350929

Chen Q, De Marco N, Yang Y et al (2015) Under the spotlight: the organic-inorganic hybrid halide perovskite for optoelectronic applications. Nano Today 10:355–396. https://doi.org/10.1016/j.nantod.2015.04.009

Chiara R, Morana M, Malavasi L (2021) Germanium-based halide perovskites: materials, properties, and applications. ChemPlusChem 86:879–888. https://doi.org/10.1002/cplu.202100191

Choudhary K et al (2015) Identification of potential replacement materials for lead in $CH_3NH_3PbI_3$ using first principle calculations. 1–12

Chung I, Lee B, He J et al (2012) All-solid-state dye-sensitized solar cells with high efficiency. Nature 485:486–489. https://doi.org/10.1038/nature11067

Cortecchia D, Dewi HA, Yin J et al (2016) Lead-free $MA_2CuCl_xBr_{4-x}$ hybrid perovskites. Inorg Chem 55:1044–1052. https://doi.org/10.1021/acs.inorgchem.5b01896

Cui XP, Jiang KJ, Huang JH et al (2015) Cupric bromide hybrid perovskite heterojunction solar cells. Synth Met 209:247–250. https://doi.org/10.1016/j.synthmet.2015.07.013

Dammak H, Yangui A, Triki S et al (2015) Structural characterization, vibrational, optical properties and DFT investigation of a new luminescent organic-inorganic material: $(C_6H_{14}N)_3Bi_2I_9$. J Lumin 161:214–220. https://doi.org/10.1016/j.jlumin.2015.01.010

Dynamics E, Doped T, Grimm J et al (2006) Light-emission and excited-state dynamics in Tm^{2+} Doped $CsCaCl_3$, $CsCaBr_3$, and $CsCaI_3$. 2093–2101

Eperon GE, Stranks SD, Menelaou C et al (2014) Formamidinium lead trihalide: a broadly tunable perovskite for efficient planar heterojunction solar cells. Energy Environ Sci 7:982–988. https://doi.org/10.1039/c3ee43822h

Fabian DM, Ardo S (2016) Hybrid organic-inorganic solar cells based on bismuth iodide and 1,6-hexanediammonium dication. J Mater Chem A 4:6837–6841. https://doi.org/10.1039/c6ta00517a

Filip MR, Giustino F et al (2016) Computational screening of homovalent lead substitution in organic-inorganic halide perovskites. J Phys Chem C 120:166–173. https://doi.org/10.1021/acs.jpcc.5b11845

Giorgi G, Yamashita K (2015) Alternative, lead-free, hybrid organic-inorganic perovskites for solar applications: a DFT analysis. Chem Lett 44:826–828. https://doi.org/10.1246/cl.150143

Han J, Nishihara S, Inoue K, Kurmoo M (2014) On the nature of the structural and magnetic phase transitions in the layered perovskite-like $(CH_3NH_3)_2[Fe^{II}Cl_4]$. Inorg Chem 53:2068–2075. https://doi.org/10.1021/ic402535u

Hao F, Stoumpos CC, Cao DH et al (2014) Lead-free solid-state organic-inorganic halide perovskite solar cells. Nat Photonics 8:489–494. https://doi.org/10.1038/nphoton.2014.82

Harikesh PC, Mulmudi HK, Ghosh B et al (2016) Rb as an alternative cation for templating inorganic lead-free perovskites for solution processed photovoltaics. Chem Mater 28:7496–7504. https://doi.org/10.1021/acs.chemmater.6b03310

Hebig JC, Kühn I, Flohre J, Kirchartz T (2016) Optoelectronic properties of $(CH_3NH_3)_3Sb_2I_9$ thin films for photovoltaic applications. ACS Energy Lett 1:309–314. https://doi.org/10.1021/acsenergylett.6b00170

Hong F, Saparov B, Meng W et al (2016) Viability of lead-free perovskites with mixed chalcogen and halogen anions for photovoltaic applications. J Phys Chem C 120:6435–6441. https://doi.org/10.1021/acs.jpcc.6b00920

Hoye RLZ, Brandt RE, Osherov A et al (2016) Methylammonium bismuth iodide as a lead-free, stable hybrid organic-inorganic solar absorber. Chem—A Eur J 22:2605–2610. https://doi.org/10.1002/chem.201505055

Hrizi C, Chaari N, Abid Y et al (2012) Structural characterization, vibrational and optical properties of a novel one-dimensional organic-inorganic hybrid based-iodobismuthate (III) material, $[C_{10}H_7NH_3]BiI_4$. Polyhedron 46:41–46. https://doi.org/10.1016/j.poly.2012.07.062

Huang C et al (2014) Novel germanium-containing perovskite material and solar cell comprising same. CN Pat. CN:23 July 2014

Huang TJ, Thiang ZX, Yin X et al (2016) $(CH_3NH_3)_2PdCl_4$: a compound with two-dimensional organic-inorganic layered perovskite structure. Chem—A Eur J 22:2146–2152. https://doi.org/10.1002/chem.201503680

Ibn-Mohammed T, Koh SCL, Reaney IM et al (2017) Perovskite solar cells: an integrated hybrid lifecycle assessment and review in comparison with other photovoltaic technologies. Renew Sustain Energy Rev 80:1321–1344. https://doi.org/10.1016/j.rser.2017.05.095

Jacobsson TJ, Pazoki M, Hagfeldt A, Edvinsson T (2015) Goldschmidt rules and strontium replacement in lead halogen perovskite solar cells: theory and preliminary experiments on $CH_3NH_3SrI_3$. J Phys Chem C 119:25673–25683. https://doi.org/10.1021/acs.jpcc.5b06436

Jiang X, Wang F, Wei Q et al (2020) Ultra-high open-circuit voltage of tin perovskite solar cells via an electron transporting layer design. Nat Commun 11:1–7. https://doi.org/10.1038/s41467-020-15078-2

Jokar E, Chien CH, Tsai CM et al (2019) Robust tin-based perovskite solar cells with hybrid organic cations to attain efficiency approaching 10%. Adv Mater 31:1–7. https://doi.org/10.1002/adma.201804835

Kaltzoglou A, Antoniadou M, Perganti D et al (2015) Mixed-halide Cs_2SnI_3Br3 perovskite as low resistance hole-transporting material in dye-sensitized solar cells. Electrochim Acta 184:466–474. https://doi.org/10.1016/j.electacta.2015.10.030

Kaltzoglou A, Antoniadou M, Kontos AG et al (2016) Optical-vibrational properties of the Cs_2SnX_6 (X = Cl, Br, I) defect perovskites and hole-transport efficiency in dye-sensitized solar cells. J Phys Chem C 120:11777–11785. https://doi.org/10.1021/acs.jpcc.6b02175

Kamminga ME, Stroppa A, Picozzi S et al (2017) Polar nature of $(CH_3NH_3)_3Bi_2I_9$ perovskite-like hybrids. Inorg Chem 56:33–41. https://doi.org/10.1021/acs.inorgchem.6b01699

Kawai T, Ishii A, Kitamuha T et al (1996) Optical absorption in band-edge region of $(CH_3NH_3)_3Bi_2I_9$ single crystals. J Phys Soc Japan 65:1464–1468. https://doi.org/10.1143/JPSJ.65.1464

Kieslich G, Sun S, Cheetham AK (2015) An extended tolerance factor approach for organic-inorganic perovskites. Chem Sci 6:3430–3433. https://doi.org/10.1039/c5sc00961h

King G, Woodward PM et al (2010) Cation ordering in perovskites. J Mater Chem 20:5785–5796. https://doi.org/10.1039/b926757c

Koh TM, Krishnamoorthy T, Yantara N et al (2015) Formamidinium tin-based perovskite with low Eg for photovoltaic applications. J Mater Chem A 3:14996–15000. https://doi.org/10.1039/c5ta00190k

Kopacic I, Friesenbichler B, Hoefler SF et al (2018) Enhanced performance of germanium halide perovskite solar cells through compositional engineering. ACS Appl Energy Mater 1:343–347. https://doi.org/10.1021/acsaem.8b00007

Kour R, Arya S, Verma S et al (2019) Potential substitutes for replacement of lead in perovskite solar cells: a review. Glob Challenges 3:1900050. https://doi.org/10.1002/gch2.201900050

Krishnamoorthy T, Ding H, Yan C et al (2015) Lead-free germanium iodide perovskite materials for photovoltaic applications. J Mater Chem A 3:23829–23832. https://doi.org/10.1039/c5ta05741h

Kumar A, Balasubramaniam KR, Kangsabanik J et al (2016) Crystal structure, stability, and optoelectronic properties of the organic-inorganic wide-band-gap perovskite $CH_3NH_3BaI_3$: Candidate for transparent conductor applications. Phys Rev B 94:1–5. https://doi.org/10.1103/PhysRevB.94.180105

Lee SJ, Shin SS, Kim YC et al (2016) Fabrication of efficient formamidinium tin iodide perovskite solar cells through SnF_2-pyrazine complex. J Am Chem Soc 138:3974–3977. https://doi.org/10.1021/jacs.6b00142

Lehner AJ, Fabini DH, Evans HA et al (2015) Crystal and electronic structures of complex bismuth iodides $A_3Bi_2I_9$ (A = K, Rb, Cs) related to perovskite: aiding the rational design of photovoltaics. Chem Mater 27:7137–7148. https://doi.org/10.1021/acs.chemmater.5b03147

Li W, Li J, Li J et al (2016) Addictive-assisted construction of all-inorganic CsSnIBr2 mesoscopic perovskite solar cells with superior thermal stability up to 473 K. J Mater Chem A 4:17104–17110. https://doi.org/10.1039/c6ta08332c

Li S, Liu P, Pan L et al (2019) The investigation of inverted p-i-n planar perovskite solar cells based on $FASnI_3$ films. Sol Energy Mater Sol Cells 199:75–82. https://doi.org/10.1016/j.solmat.2019.04.023

Li X, Kepenekian M, Li L et al (2022) Tolerance factor for stabilizing 3d hybrid halide perovskitoids using linear diammonium cations. J Am Chem Soc 144:3902–3912. https://doi.org/10.1021/jacs.1c11803

Liao W, Zhao D, Yu Y et al (2016) Lead-free inverted planar formamidinium tin triiodide perovskite solar cells achieving power conversion efficiencies up to 6.22%. Adv Mater 28:9333–9340. https://doi.org/10.1002/adma.201602992

Liao Y, Liu H, Zhou W et al (2017) Highly oriented low-dimensional tin halide perovskites with enhanced stability and photovoltaic performance. J Am Chem Soc 139:6693–6699. https://doi.org/10.1021/jacs.7b01815

Lim EL, Hagfeldt A, Bi D (2021) Toward highly efficient and stable Sn^{2+} and mixed Pb^{2+}/Sn^{2+} based halide perovskite solar cells through device engineering. Energy Environ Sci 14:3256–3300. https://doi.org/10.1039/d0ee03368e

Liu X, Yan K, Tan D et al (2018) Solvent engineering improves efficiency of lead-free tin-based hybrid perovskite solar cells beyond 9%. ACS Energy Lett 3:2701–2707. https://doi.org/10.1021/acsenergylett.8b01588

Liu XJ, Matsuda K, Moritomo Y et al (1999) Electronic structure of the gold complexes $Cs_2Au_2X_6$ (X=I, Br, and Cl). Phys Rev B 59:7925–7930. https://doi.org/10.1103/PhysRevB.59.7925

Mostafa Mas MF, Maa (1977) Field dependence of the susceptibility maximum for two-dimensional antiferromagnet. Phys Lett A 61:183–184. https://doi.org/10.1016/0375-9601(77)90287-0

Macdonald TJ, Lanzetta L, Liang X et al (2022) Engineering stable lead-free tin halide perovskite solar cells: lessons from materials chemistry. Adv Mater 2206684:2206684. https://doi.org/10.1002/adma.202206684

Manser JS, Christians JA, Kamat PV (2016) Intriguing optoelectronic properties of metal halide perovskites. Chem Rev 116:12956–13008. https://doi.org/10.1021/acs.chemrev.6b00136

Michaelson HB et al (1977) The work function of the elements and its periodicity. J Appl Phys 48:4729–4733. https://doi.org/10.1063/1.323539

Mitzi DB, Liang K (1997) Preparation and properties of $(C_4H_9NH_3)_2EuI_4$: a luminescent organic-inorganic perovskite with a divalent rare-earth metal halide framework. Chem Mater 9:2990–2995. https://doi.org/10.1021/cm970352d

Morss LR, Siegal M, Stenger L, Edelstein N (1970) Preparation of cubic chloro complex compounds of trivalent metals: Cs_2NaMCl_{61} a. Inorg Chem 9:1771–1775. https://doi.org/10.1021/ic50089a034

Mostafa MF, Willett RD et al (1971) Magnetic properties of ferrous chloride complexes with two-dimensional structures. Phys Rev B 4:2213–2215. https://doi.org/10.1103/PhysRevB.4.2213

Öz S, Hebig JC, Jung E et al (2016) Zero-dimensional $(CH_3NH_3)_3Bi_2I_9$ perovskite for optoelectronic applications. Sol Energy Mater Sol Cells 158:195–201. https://doi.org/10.1016/j.solmat.2016.01.035

Park BW, Philippe B, Zhang X et al (2015) Bismuth based hybrid perovskites $A_3Bi_2I_9$ (A: Methylammonium or Cesium) for solar cell application. Adv Mater 27:6806–6813. https://doi.org/10.1002/adma.201501978

Pazoki M, Jacobsson TJ, Hagfeldt A et al (2016) Effect of metal cation replacement on the electronic structure of metalorganic halide perovskites: replacement of lead with alkaline-earth metals. Phys Rev B 93:1–10. https://doi.org/10.1103/PhysRevB.93.144105

Perera S, Hui H, Zhao C et al (2016) Chalcogenide perovskites—an emerging class of ionic semiconductors. Nano Energy 22:129–135. https://doi.org/10.1016/j.nanoen.2016.02.020

Peresh EY, Sidei VI, Zubaka OV, Stercho IP (2011) K2(Rb2, Cs2, Tl2)TeBr 6(I6) and Rb3(Cs3)Sb2(Bi2)Br9(I9) perovskite compounds. Inorg Mater 47:208–212. https://doi.org/10.1134/S0020168511010109

Qian J, Xu B, Tian W (2016) A comprehensive theoretical study of halide perovskites ABX_3. Org Electron 37:61–73. https://doi.org/10.1016/j.orgel.2016.05.046

Retuerto M, Emge T, Hadermann J et al (2013) Synthesis and properties of charge-ordered Thallium Halide perovskites, $CsTl^{+}{}_{0.5}Tl^{3+0.5}X_3$ (X = F or Cl): Theoretical precursors for superconductivity? Chem Mater 25(20): 4071–4079. https://doi.org/10.1021/cm402423x

Riggs SC, Shapiro MC, Corredor F et al (2012) Single crystal growth by self-flux method of the mixed valence gold halides $Cs_2[Au^{I}X_2][Au^{III}X_4]$ (X=Br, I). J Cryst Growth 355:13–16. https://doi.org/10.1016/j.jcrysgro.2012.06.039

Saliba M et al (2016) Incorporation of rubidium cationsinto perovskite solar cells improvesphotovoltaic performance. Science (80-) 354:203–206. https://doi.org/10.1126/science.aah5557

Sani F, Shafie S, Lim HN, Musa AO (2018) Advancement on lead-free organic-inorganic halide perovskite solar cells: a review. Materials (Basel) 11:1–17. https://doi.org/10.3390/ma11061008

Saparov B, Hong F, Sun JP et al (2015) Thin-film preparation and characterization of $Cs_3Sb_2I_9$: a lead-free layered perovskite semiconductor. Chem Mater 27:5622–5632. https://doi.org/10.1021/acs.chemmater.5b01989

Saparov B, Mitzi DB et al (2016) Organic-inorganic perovskites: structural versatility for functional materials design. Chem Rev 116:4558–4596. https://doi.org/10.1021/acs.chemrev.5b00715

Shannon RD et al (1976) Revised effective ionic radii and systematic studies of interatomic distances in halides and chalcogenides. Acta Crystallogr Sect B Struct Sci A32:751–767. https://doi.org/10.1107/S0567739476001551

Slavney AH, Hu T, Lindenberg AM, Karunadasa HI (2016) A bismuth-halide double perovskite with long carrier recombination lifetime for photovoltaic applications. J Am Chem Soc 138:2138–2141. https://doi.org/10.1021/jacs.5b13294

Stoumpos CC, Frazer L, Clark DJ et al (2015) Hybrid germanium iodide perovskite semiconductors: active lone pairs, structural distortions, direct and indirect energy gaps, and strong nonlinear optical properties. J Am Chem Soc 137:6804–6819. https://doi.org/10.1021/jacs.5b01025

Stranks SD, Eperon GE, Grancini G et al (2013) Electron-hole diffusion lengths exceeding 1 micrometer in an organometal trihalide perovskite absorber. Science (80-) 342:341–344. https://doi.org/10.1126/science.1243982

Sun PP, Li QS, Yang LN, Li ZS (2016a) Theoretical insights into a potential lead-free hybrid perovskite: substituting Pb^{2+} with Ge^{2+}. Nanoscale 8:1503–1512. https://doi.org/10.1039/c5nr05337d

Sun S, Tominaka S, Lee JH et al (2016b) Synthesis, crystal structure, and properties of a perovskite-related bismuth phase, $(NH_4)_3Bi_2I_9$. APL Mater 4. https://doi.org/10.1063/1.4943680

Sun YY, Shi J, Lian J et al (2016c) Discovering lead-free perovskite solar materials with a split-anion approach. Nanoscale 8:6284–6289. https://doi.org/10.1039/c5nr04310g

Suta M, Larsen P, Lavoie-Cardinal F, Wickleder C (2014) Photoluminescence of $CsMBr3:E^{u2}+$ (M=Mg, Ca, Sr)—a novel strategy for the development of low-energy emitting phosphors. J Lumin 149:35–44. https://doi.org/10.1016/j.jlumin.2013.12.040

Suta M, Wickleder C et al (2015) Photoluminescence of $CsMI_3:Eu^{2+}$ (M = Mg, Ca, and Sr)-a spectroscopic probe on structural distortions. J Mater Chem C 3:5233–5245. https://doi.org/10.1039/c5tc00515a

Suta M, Urland W, Daul C, Wickleder C (2016) Photoluminescence properties of Yb^{2+} ions doped in the perovskites $CsCaX_3$ and $CsSrX_3$ (X = Cl, Br, and I)-a comparative study. Phys Chem Chem Phys 18:13196–13208. https://doi.org/10.1039/c6cp00085a

Tiwari D, Hutter OS, Longo G (2021) Chalcogenide perovskites for photovoltaics: current status and prospects. J Phys Energy 3. https://doi.org/10.1088/2515-7655/abf41c

Travis W, Glover ENK, Bronstein H et al (2016) On the application of the tolerance factor to inorganic and hybrid halide perovskites: a revised system. Chem Sci 7:4548–4556. https://doi.org/10.1039/c5sc04845a

Turkevych I, Kazaoui S, Ito E et al (2017) Photovoltaic rudorffites: lead-free silver bismuth halides alternative to hybrid lead halide perovskites. Chemsuschem 10:3754–3759. https://doi.org/10.1002/cssc.201700980

Uribe JI, Ramirez D, Osorio-Guillén JM et al (2016) $CH_3NH_3CaI_3$ perovskite: synthesis, characterization, and first-principles studies. J Phys Chem C 120:16393–16398. https://doi.org/10.1021/acs.jpcc.6b04207

Volonakis G, Filip MR, Haghighirad AA et al (2016) Lead-free halide double perovskites via heterovalent substitution of noble metals. J Phys Chem Lett 7:1254–1259. https://doi.org/10.1021/acs.jpclett.6b00376

Wang N, Zhou Y, Ju MG et al (2016) Heterojunction-depleted lead-free perovskite solar cells with coarse-grained β-γ-$CsSnI_3$ thin films. Adv Energy Mater 6:1–10. https://doi.org/10.1002/aenm.201601130

Wang F, Jiang X, Chen H et al (2018) 2D-Quasi-2D-3D hierarchy structure for tin perovskite solar cells with enhanced efficiency and stability. Joule 2:2732–2743. https://doi.org/10.1016/j.joule.2018.09.012

Wei F, Deng Z, Sun S et al (2017) Synthesis and properties of a lead-free hybrid double perovskite: $(CH_3NH_3)_2$AgBiBr6. Chem Mater 29:1089–1094. https://doi.org/10.1021/acs.chemmater.6b03944

Yamada K, Sera H, Sawada S et al (1997) Reconstructive phase transformation and kinetics of $Cs_3Sb_2I_9$ by means of Rietveld analysis of x-ray diffraction and127i NQR. J Solid State Chem 134:319–325. https://doi.org/10.1006/jssc.1997.7562

Yang L, Ding P, Luo Y et al (2019) Exploration of catalytic nucleic acids on porphyrin metalation and peroxidase activity by in vitro selection of aptamers for N-methyl mesoporphyrin IX. ACS Comb Sci 21:83–89. https://doi.org/10.1021/acscombsci.8b00129

Ye HY, Zhang Y, Fu DW, Xiong RG (2014) A displacive-type metal crown ether ferroelectric compound: Ca($NO_3$2(15-crown-5). Angew Chemie—Int Ed 53:6724–6729. https://doi.org/10.1002/anie.201402339

Yin ZP, Kotliar G et al (2013) Rational material design of mixed-valent high-Tc superconductors. Epl 101. https://doi.org/10.1209/0295-5075/101/27002

Zaleski J, Pietraszko A et al (1996) Structure at 200 and 298 K and X-ray investigations of the phase transition at 242 K of $[NH_2(CH_3)_2]_3Sb_2Cl_9$ (DMACA). Acta Crystallogr Sect B Struct Sci 52:287–295. https://doi.org/10.1107/S0108768195010615

Zhang M, Lyu M, Yun JH et al (2016a) Low-temperature processed solar cells with formamidinium tin halide perovskite/fullerene heterojunctions. Nano Res 9:1570–1577. https://doi.org/10.1007/s12274-016-1051-8

Zhang X, Wu G, Gu Z et al (2016b) Active-layer evolution and efficiency improvement of $(CH_3NH_3)_3Bi_2I_9$-based solar cell on TiO_2-deposited ITO substrate. Nano Res 9:2921–2930. https://doi.org/10.1007/s12274-016-1177-8

Zhang Q, Hao F, Li J et al (2018) Perovskite solar cells: must lead be replaced–and can it be done? Sci Technol Adv Mater 19:425–442. https://doi.org/10.1080/14686996.2018.1460176

Zhao Z et al (2017) Mixed-organic-cation tin iodide for lead-free perovskite solar cells with an efficiency of 8.12%. Adv Sci 4:1700204. https://doi.org/10.1002/advs.201700204

Chapter 5
Pioneering the Path: Unveiling Exciting Applications for Integrated Perovskite Solar Cells

Abstract Metal halide perovskites, particularly perovskite solar cells (PSCs), exhibit outstanding properties, propelling rapid progress in photovoltaic (PV) devices. PSCs are crucial for transformative applications like tandem PV cells, space solutions, PV-integrated energy storage, catalysis, and building-integrated photovoltaics (BIPVs). Tandem solar cells face challenges in optimizing low-bandgap perovskites and improving the efficiency of wide-bandgap materials. PSCs offer high output voltage, enabling integrated energy storage with carbon-based materials. Challenges include matching photocurrents, operational stability, and CO_2 reduction. Semi-transparent PSCs for BIPVs show promise but need efficiency and transparency improvements. Addressing these challenges in materials and system integration will advance and commercialize perovskite-based solar cells across applications, driving innovation in renewable energy technologies.

Keywords Building integrated photovoltaics · Tandem solar cells · PV-integrated energy storage · Smart PV-windows · PVs for space applications · PV-catalysis

5.1 Introduction

The potential for low cost, excellent efficiency, and solution processability has attracted significant attention to organic/inorganic metal halide perovskites as crucial materials for future photovoltaic technologies. The unique characteristics of perovskites and the swift progress in solar cell efficiency have made it possible for them to be incorporated into various practical applications, such as building-integrated photovoltaics, tandem solar cells, space applications, photovoltaic-driven catalysis, and their association with batteries and supercapacitors for energy storage systems.

In recent years, metal halide perovskites have gained attention as promising materials for photovoltaics because of their exceptional optical and electrical properties. The chemical structure of these materials, denoted as ABX_3 involves MA^+, FA^+, or Cs^+ as A, and Pb^{2+} or Sn^{2+} as B, with I^-, Br^-, Cl^-, or combinations of those

A. Srivastava and P. M. Shirage, *Perovskite Solar Cells*, Springer Tracts in Electrical and Electronics Engineering, https://doi.org/10.1007/978-981-97-7624-5_5

as X. These materials possess outstanding charge-carrier mobilities, low exciton binding energies, tunable bandgaps, ambipolar transport, extended carrier lifetimes, long diffusion lengths, and high defect tolerance (Zhang et al. 2021a). The impressive properties of perovskites have enabled the fast growth of photovoltaic devices that utilize these materials, with power conversion efficiencies (PCEs) increasing from 3.8 to 25.7% (NREL 2023). PSCs offer various benefits in comparison to other existing and growing PV technologies. For example, traditional silicon solar cells need pure silicon, obtained when heated with sand at high temperatures (> 1000 °C), which results in intricate synthesis techniques and greenhouse gas emissions, all contributing to high production costs. In contrast, perovskites can be manufactured using low-embedded energy through solution processing with commercial coating technologies. Compared to other emerging PV technologies like organic solar cells (OSCs), PSCs have a higher relative dielectric constant, a more effective charge-carrier mobility, and a more considerable charge-carrier diffusion length at open circuits (Zhang et al. 2021a).

In response to the remarkable progress in the PCEs of PSCs, an innovative and growing field of research has emerged that aims to utilize PSCs as an energy source for integrated systems like energy conversion and storage devices. Even though they are still in their early stages, efforts to incorporate PSCs into an extensive range of applications have displayed notable potential. The integration of advanced PSCs into a diverse array of next-generation applications is covered in this section, including tandem solar cells, building-integrated PVs (BIPVs), space applications, batteries powered by PV, supercapacitors, and energy sources for the catalytic synthesis of high-value chemicals (as shown in Fig. 5.1). Lastly, we briefly forecast this thriving research domain's difficulties and future prospects.

5.2 Tandem Solar Cells (TSCs)

Significant development has been made in increasing the PCEs of single-junction PSCs, which have reached close to their theoretical limit of 26.1% under one Sun irradiation. In order to improve the PCEs and approach the Shockley-Queisser limit of ~33%, reducing photon transmission losses and thermalization of high-energy carriers is essential. Combining a top-layer wide-bandgap absorber with a bottom-layer low-bandgap absorber in tandem solar cells is one way to reduce these losses and exceed the limit for a single-junction device. This design allows the absorption of high-energy photons by the top layer while low-energy photons pass through to the bottom layer to be harvested. Two types of tandem solar cell structures exist: the 2-terminal (2T) and 4-terminal (4T) structures, as illustrated in Fig. 5.2 (Zhu et al. 2021). The 2T configuration involves stacking both sub-cells on a single substrate, while the 4T configuration involves fabricating two separate cells that are physically connected. The 2T arrangement is preferred for real-life applications because of its cheaper manufacturing cost, use of only two electrodes, and reduced optical loss resulting from the dearth of a physical gap between the sub-cells. According to theoretical analysis, stacked cell designs with two- and three-junction architectures

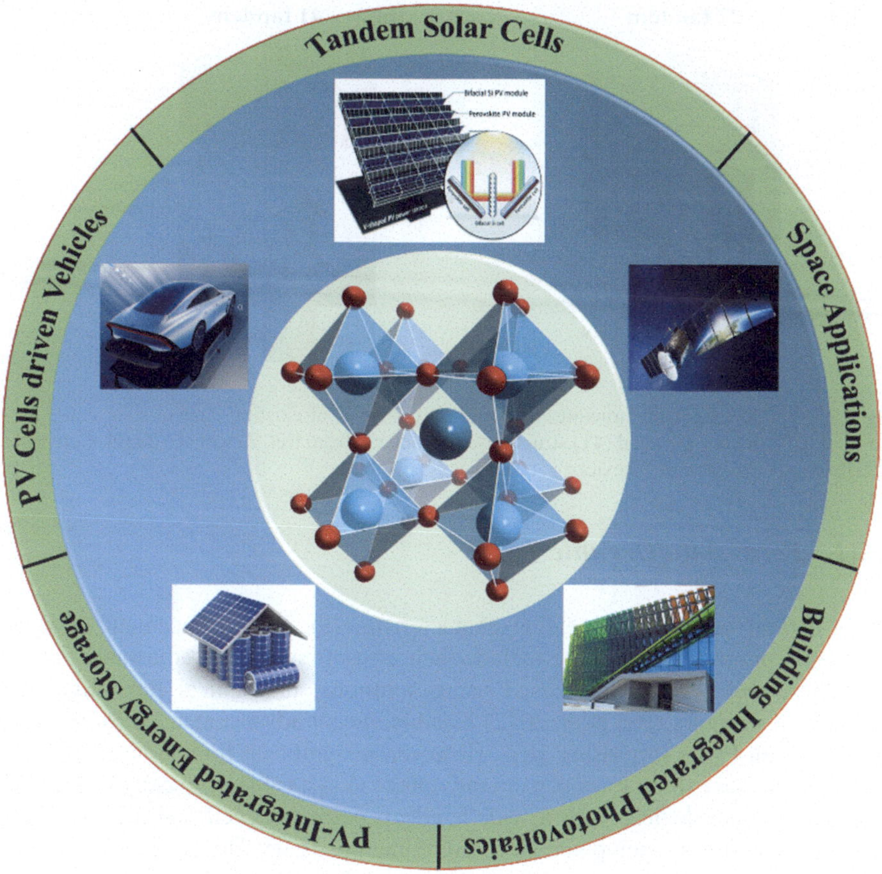

Fig. 5.1 Schematic illustration of cutting-edge PSCs into a diverse array of next-generation applications, such as tandem solar cells, building-integrated PVs (BIPVs), space applications, batteries powered by PV, supercapacitors, and energy sources for catalytic synthesis of high-value chemicals, etc.

may correspondingly attain PCEs of up to 42% and 49%. The maximum efficiency might be raised to 68% and 86% under concentrated and un-concentrated sunlight by stacking an infinite number of solar cells (De Vos et al. 1980).

Considering the standpoint, the expense associated with synthesizing multi-junction stacked devices experiences a substantial increase, potentially surpassing the efficiency benefits. The ensuing subsections will provide a comprehensive discussion of the notable strides achieved in perovskite-based tandem solar cells.

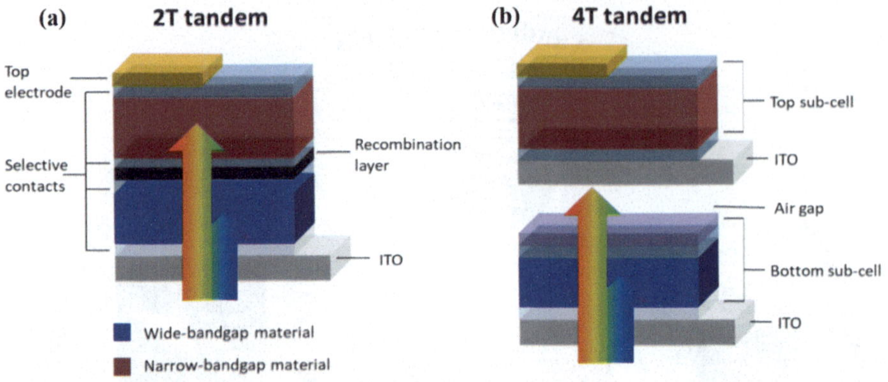

Fig. 5.2 Typical schematic structures of two different types of existing tandem solar cells. **a** 2-terminal (2T) and **b** 4-terminal (4T) structures. Reproduced from Ref. Zhu et al. (2021), Copyright (2021) with permission of Elsevier

5.2.1 *Perovskite/Organic TSCs*

OSCs exhibit numerous appealing characteristics for the advancement of next-generation photovoltaics. These include their cost-effectiveness, adjustable optical properties, ability to be processed using solutions, mechanical flexibility, and lightweight structure (Pan et al. 2022) Notably, recent advancements have resulted in OSCs with PCEs surpassing 18%. However, a significant limitation of OSCs is their tendency to exhibit low open circuit voltage (V_{OC}) values, typically below 1 V. Utilizing tandem devices with a wide-bandgap perovskite cell and a low optical gap organic cell has much potential to overcome this issue. This approach not only enhances the overall device performance but also leverages the hydrophobic properties of the organic layers to potentially stabilize the perovskite sub-cell, which is particularly susceptible to moisture-induced degradation.

Yang et al. pioneered the development of tandem devices combining perovskite and organic materials, achieving noteworthy results. Specifically, they successfully fabricated tandem devices with a perovskite absorber, $MAPbI_3$, and an organic semiconductor absorber, namely an IR-sensitive block copolymer $PBSeDTEG_8$:fullerene mixture. The tandem devices displayed a PCE of 10.2% and V_{OC} of 1.52 V, demonstrating the viability of this approach for high-performance photovoltaic systems (Chen et al. 2015). In their research, Liu et al. explored an alternative approach for tandem devices by inverting the layer order within the structure. They employed an organic layer as the bottom sub-cell and a thin layer of $MAPbI_3$ (90 nm) as the top sub-cell. Remarkably, this configuration resulted in tandem devices with a PCE of 16% and V_{OC} of 1.63 V. The PCE of these tandem devices surpassed that of a single-junction perovskite device with an equivalent perovskite layer thickness (9.1%) and a single-junction organic device (9.7%) (Liu et al. 2016).

In a separate study, Yang et al. utilized a perovskite composition, $FA_{0.8}MA_{0.02}Cs_{0.18}PbI_{1.8}Br_{1.2}$, that has a wide-bandgap of 1.77 eV as the top sub-cell, coupled with a PBDBT-2F: Y_6:$PC_{71}BM$ mixture featuring a small optical gap of 1.41 eV as the bottom sub-cell. This configuration led to the fabrication of 2T perovskite/organic TSC, achieving a remarkable PCE of 20.6% and V_{OC} of 1.90 V (Fig. 5.3a, b) (Chen et al. 2020). These findings demonstrate the potential of tandem devices to enhance the overall performance of photovoltaic systems. The study also revealed an additional benefit of the perovskite sub-cell in the tandem device, wherein it served as a UV filter and effectively eliminated the UV sensitivity of the organic sub-cell. This UV filtering capability resulted in improved photostability of the tandem device. The perovskite sub-cell acted as a protective layer, shielding the organic sub-cell from the potentially damaging effects of UV radiation. Consequently, the tandem device exhibited enhanced stability under prolonged exposure to light, highlighting the perovskite sub-cell's advantageous role in preserving the device's overall performance (Chen et al. 2020).

Wang et al. demonstrated the crucial role of HTM in the interconnecting layers for the development of monolithic perovskite/organic TSC. The HTM mitigated charge accumulation at the interface, thereby minimizing voltage loss. To investigate this effect, the researchers used a narrow-optical gap PM_6:Y_6-BO mixture as

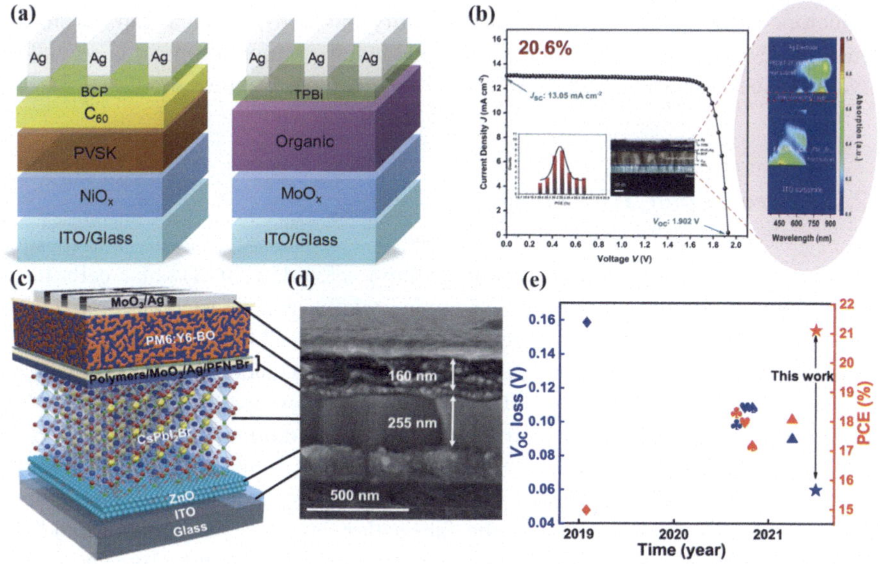

Fig. 5.3 **a** Typical schematics of single-junction tandem solar cells, **b** J–V analysis of a Perovskite-Organic TSCs having active area 0.062 cm^2, Reproduced from Ref. Chen et al. (2020), Copyright (2020) with permission of Elsevier. **c, d** schematics and cross-sectional FESEM image of another perovskite-organic TSC consisting of PM_6:Y_6-BO blend as the bottom sub-cell, and **e** device performance and stability analysis, Reproduced from Ref. Wang et al. (2021), Copyright (2021) with permission of the American Chemical Society

the bottom sub-cell and a wide-bandgap $CsPbI_2Br$ material as the top sub-cell. A 4-butyl-N,N-diphenylaniline homopolymer (polyTPD) was chosen as the HTL in the interconnecting layer. By implementing this configuration, the tandem device achieved a notable PCE of 21.1% and V_{OC} of 1.96 V, as shown in Fig. 5.3c–e. These findings underscore the significance of incorporating an HTM in the interconnecting layers to minimize charge accumulation, reduce voltage loss, and improve the overall efficacy of monolithic perovskite/organic tandem solar cells (Wang et al. 2021).

The development of non-fullerene acceptors (NFAs) in OSC technology has brought about notable advancements due to their convenient synthetic routes and ability to adjust energy levels. These NFAs have significantly lowered V_{OC} loss in OSCs, enhancing device performance. In fact, recent advances in developing new NFAs have enabled single-junction OSCs to achieve a certified PCE of 19.2%. This significant performance improvement is highly advantageous for the development of perovskite/organic tandem devices that are more effective. In a recent study by Brinkmann et al., they demonstrated perovskite–organic tandem solar cells featuring a certified PCE of 23.1% and a notably high V_{OC} of 2.15 V. This remarkable achievement was based on a ternary system incorporating a Y_6-containing organic semiconductor as the OSC component (Brinkmann et al. 2022).

Looking ahead, further advancements in OSC sub-cells are expected to continue to drive the development of more efficient tandem devices. Future research on perovskite/organic tandem solar cells should concentrate on developing organic semiconductors with excellent environmental stability and narrow -optical gap. To ensure increased device performance and stability in tandem configurations, it is crucial to solve the issue of phase segregation in wide-bandgap perovskites.

5.2.2 Perovskite/CIGS TCS

Polycrystalline thin-film copper indium gallium selenide (CIGS) solar cells have proven themselves to be a mature and commercially available technology. These solar cells exhibit a noteworthy record efficiency of 23.4% for single-junction devices, which enhances their appeal for integration into tandem solar cells made of perovskites. The direct bandgap of CIGS-based absorbers may be tuned to a value of about 1 eV, which is one of their key advantages. Additionally, CIGS materials demonstrate a higher absorption coefficient of about 10^5 cm^{-1}. This high absorption coefficient substantially reduces absorber thickness requirements for tandem devices, leading to cost savings in the fabrication process. The tunable bandgap and high absorption coefficient properties of CIGS make it an attractive choice for incorporation into tandem solar cells made of perovskites, offering the potential for enhanced overall performance and cost-effectiveness in producing these devices (Yakushev et al. 2004; Barkhouse et al. 2015).

Albrecht et al. conducted a study investigating the use of dual p-type HTLs comprising NiO_x that is being deposited in atomic layers and spin-coated PTAA on a rough CIGS bottom sub-cell (Fig. 5.4a, b). The deposition of these HTLs served two purposes: preventing device shunting and improving the performance of monolithic perovskite/CIGS TSCs. Through this approach, they achieved a PCE of 21.6% for a device with an active area of 0.78 cm^2 (Fig. 5.4c) (Jošt et al. 2019). However, despite the promising PCE, there was a significant mismatch among the two sub-cells in the reported photocurrent. This mismatch was ascribed to parasitic absorption and the rough interfaces present in the device (Jošt et al. 2019). In a subsequent study in 2019, the same research group demonstrated the efficacy of utilizing a SAM that selectively binds to the oxide layer of the rough CIGS surface. This approach led to a PCE improvement in monolithic perovskite/CIGS TSCs. The devices achieved a remarkable PCE of 23.3% when a greater active area of 1.03 cm^2 is being used (Fig. 5.4d–e) (Al-Ashouri et al. 2019). It was hypothesized that the SAM formed an energetically favorable interface with the perovskite, serving as an effective hole-selective contact without incorporating non-radiative losses. These findings highlight the significance of optimizing the interface and surface properties in monolithic perovskite/CIGS tandem solar cells to improve overall performance and achieve higher efficiencies.

Jošt et al. recently performed a study and presented their findings on monolithic perovskite/CIGS TSCs. These devices achieved a certified PCE of 24.2%. This impressive performance was accomplished by incorporating an extensive bandgap perovskite material (with a bandgap energy of 1.68 eV) that contained a PEAI additive (Jošt et al. 2022). Additionally, the devices employed a Me-4PACz monolayer as the HTM and a LiF interlayer. Utilizing these materials and interfaces demonstrated the considerable potential of perovskite/CIGS TSCs. The incorporation of a large bandgap perovskite, along with the PEAI additive, contributed to enhanced device performance. The Me-4PACz monolayer was an effective HTM, facilitating charge transport within the device. Furthermore, including a LiF interlayer helped optimize the interface properties, further improving the overall device efficiency. These findings highlight the promise and prospects of perovskite/CIGS TSCs, showcasing the remarkable achievements in reaching a certified PCE of 24.2%. Continued advancements in material design and interface engineering hold great potential for further enhancing the performance and efficiency of such tandem solar cell configurations.

5.2.3 *Perovskite/Perovskite TSCs*

'All-perovskite tandem solar cells (all-PTSCs) present an appealing option for solar energy conversion despite the existing challenges that necessitate attention and resolution. In the structure of an all-PTSC, two types of perovskite materials are employed: a wide-bandgap perovskite having a bandgap energy of approximately 1.7 eV for the top sub-cell and a narrow-bandgap perovskite having a bandgap energy of around 1 eV for the bottom sub-cell. In the context of single junction PSCs, it is customary to fabricate them with a bandgap ranging from 1.5 to 1.7 eV, exemplified by the widely used $MAPbI_3$ perovskite material. This range of bandgap

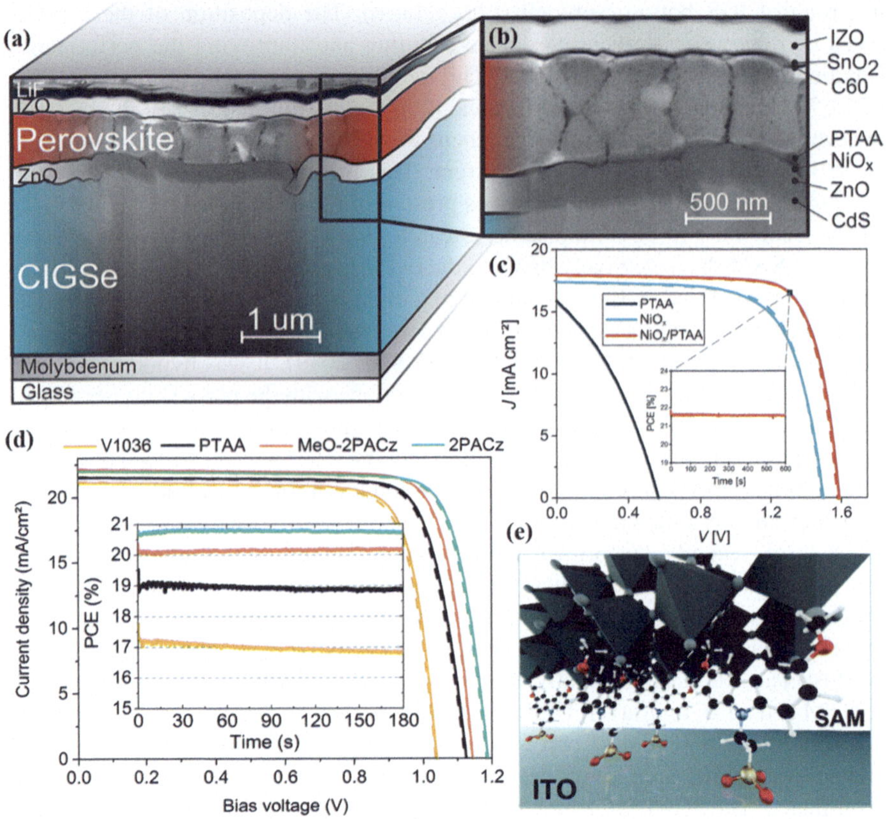

Fig. 5.4 **a–c** Schematics, cross-section, and performance of Perovskite/CIGS TSC of dual p-type HTLs consisting of atomic layer deposited NiO_x and spin-coated PTAA on a rough CIGS bottom sub-cell, Reproduced from Ref. Jošt et al. (2019), Copyright (2019) with permission of the American Chemical Society. **d–e** Schematic and understanding of the performance of monolithic perovskite/CIGS TSCs by utilizing self-assembled monolayer (SAM), Reproduced from Ref. Al-Ashouri et al. (2019), Copyright (2019) with permission of the Royal Society of Chemistry

energies aligns with the requirement for the wide-bandgap sub-cell in all-PTSCs. To achieve efficient all-PTSCs, it is imperative to develop perovskite materials with suitable bandgap energies, satisfying the specific requirements for both the wide-bandgap and narrow-bandgap sub-cells. These advancements in material design and synthesis hold promise for addressing the challenges associated with all-PTSCs and unlocking their full potential in solar energy conversion (Yang et al. 2019).

It is important to note that fabricating 2-T all-PTSCs through solution processes poses a significant challenge. The challenge arises from the fact that since the materials are frequently soluble in the same processing solvents, depositing the top sub-cell could damage or dissolve the bottom sub-cell. To address this issue, implementing an interconnecting layer with orthogonal solubility between the sub-cells can be highly effective in resolving this problem and protecting the bottom perovskite layer. Palmstrom et al. in 2019 presented an efficient surface treatment strategy for the interconnecting layer using a thin layer (1 nm) of poly(ethylenimine) ethoxylated (PEIE) and an aluminum zinc oxide (AZO) film deposited through ALD (Palmstrom et al. 2019). This surface treatment approach provides several benefits to the tandem device structure. Firstly, the PEIE layer acts as a protective barrier, preventing the dissolution or damage of the underlying bottom perovskite layer during the deposition of the top sub-cell. The PEIE layer is selected due to its compatibility with the processing solvents utilized for the top sub-cell materials, while offering limited solubility in those solvents. This characteristic allows for the deposition of the top sub-cell without compromising the integrity of the bottom perovskite layer. Secondly, the ALD-deposited AZO film serves as a transparent conducting layer that facilitates efficient charge extraction from the top sub-cell. This layer ensures proper electrical connectivity between the sub-cells, enabling the series connection required for the tandem cell operation.

By incorporating this surface treatment strategy, Palmstrom et al. demonstrated an effective solution to mitigate the dissolution and damage issues that can occur during the fabrication of 2-terminal all-PTSCs (Fig. 5.5a). This advancement contributes to developing reliable and efficient tandem solar cell devices. Using this surface treatment strategy, researchers achieved the fabrication of 2-terminal all-PTSCs with impressive PCEs. Specifically, PCEs of 23.1% (Fig. 5.5b) and 21.3% (Fig. 5.5c) were achieved for rigid and flexible substrates accordingly. This outcome is noteworthy as it demonstrates the successful formation of highly efficient all-PTSCs and showcases their ability to possess desirable characteristics such as light-weightness and flexibility (Palmstrom et al. 2019). Moreover, the ability of these all-PTSCs to exhibit a lightweight and flexible form factor is significant for practical applications. The lightweight nature of the devices makes them suitable for various applications where weight reduction is essential, such as portable electronics or building-integrated photovoltaics. Additionally, the flexible nature of the devices allows for their integration into curved or irregular surfaces, expanding the range of potential deployment scenarios.

To mitigate losses arising from the limited carrier diffusion length in comparison to the thickness of the perovskite film, a passivation technique was employed on Pb–Sn perovskite materials using 4-trifluoromethylphenylammonium (CF3-PA). The carrier diffusion length increased significantly during this passivation process, exceeding 5 μm. By implementing this approach, the researchers successfully designed all-perovskite tandem solar cells, achieving a certified PCE of 26.4%. Remarkably, these cells retained over 90% of the original PCE even after 600 h of continuous irradiation at the MPP under ambient conditions. It is noteworthy to mention that a recent development involved the demonstration of a monolithic perovskite-perovskite-silicon triple-junction TSC, exhibiting an efficiency of over 20%, a V_{OC} of 2.74 V, and a FF of 86% (Fig. 5.5d–f) (Zheng et al. 2022). However, considering the increased complexity and cost associated with such a sophisticated device structure, further improvements in PCEs would be required to justify the additional expenses involved.

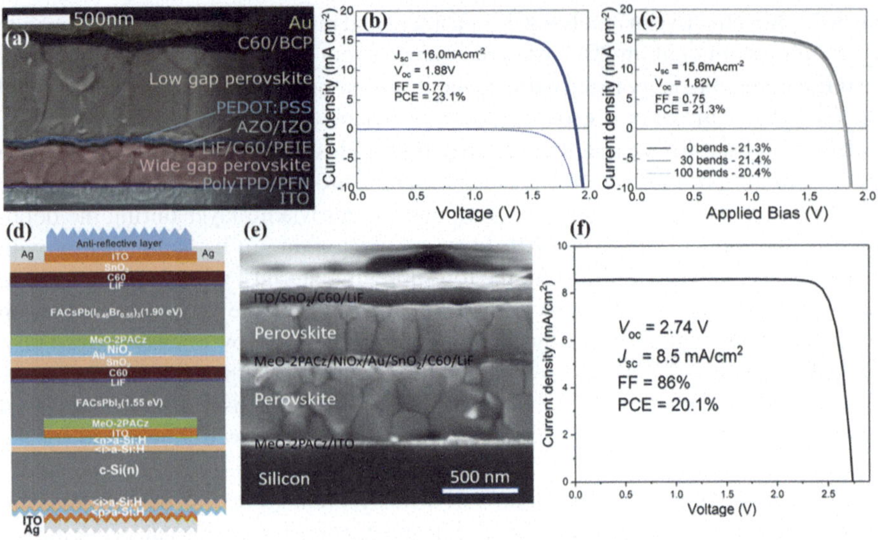

Fig. 5.5 **a–c** Cross-sectional FESEM image of AZO/IZO deposited perovskite/perovskite TSCs and their respective performances, Reproduced from Ref. Palmstrom et al. (2019), Copyright (2019) with permission of Elsevier. **d–f** schematics, cross-sectional FESEM image, and respective *J–V* analysis of interfacial passivated perovskite/perovskite TSCs using CF3-PA exhibits an efficiency of over 20%, a V_{OC} of 2.74 V, and fill factor (FF) of 86%, Reproduced from Ref. Zheng et al. (2022), Copyright (2022) with permission of the American Chemical Society

5.2.4 *Perovskite/Si TSCs (PSTSCs)*

Silicon (Si) solar cells, boasting a dominant market share of over 90%, are highly mature candidates for integration with PSCs in tandem device configurations. Si solar cells offer several advantages, including a low fabrication cost, appropriate bandgap, exceptional performance, and a lifespan exceeding 20 years. Leveraging the integration of PSCs and Si cells in tandem devices presents a significant potential to achieve high-efficiency PV systems. However, the development of PSTSCs poses certain challenges (Shen et al. 2020). One of these challenges involves the need for transparent and conductive electrodes that transmit near-infrared (NIR) light through the semi-transparent perovskite top sub-cell to reach the underlying Si sub-cell. This requirement arises from the desire to efficiently utilize the NIR portion of the incident light spectrum, which contains valuable energy for electricity generation. The transparency and conductivity of the electrodes are crucial for facilitating the passage of NIR light while ensuring proper electrical functionality of the tandem device. Addressing this challenge involves the exploration and development of suitable electrode materials and structures that strike a balance between transparency and conductivity. Effective electrode design and optimization enable the transmittance of NIR light, thereby maximizing the overall PCE of the PSTSCs. By overcoming this obstacle, researchers aim to unlock the full potential of PSTSCs and establish them as promising candidates for high-efficiency PV systems.

Wang et al., in a recent study, used a thermally evaporated semi-transparent electrode consisting of a multilayer structure of MoO_3/Au/MoO_3 was utilized for the perovskite sub-cell in a TSC configuration. This electrode design aimed to achieve the transparency and conductivity necessary for efficient light transmission and charge extraction in the device. The researchers successfully fabricated a champion 4T PSTSCs by employing this transparent conducting electrode. Notably, the tandem device exhibited a PCE of 27%, surpassing the efficiencies of the individual sub-cells. This improvement in PCE signifies the enhanced performance achieved through the synergistic operation of the silicon and perovskite sub-cells in the tandem configuration (Fig. 5.6a, b) (Wang et al. 2020).

The interconnection layer (ICL) plays a vital role within the 2T TSC configuration by serving as an optical and electrical link between the top and bottom sub-cells. Its

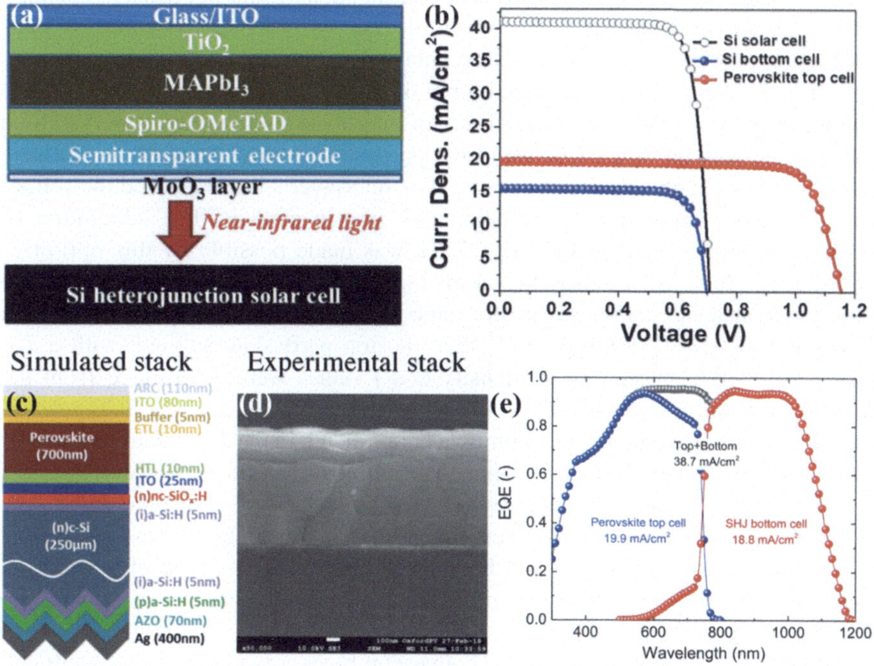

Fig. 5.6 **a**, **b** Performance improvement through the synergistic operation of the perovskite and silicon sub-cells in the tandem configuration, Reproduced from Ref. Wang et al. (2020), Copyright (2019) with permission of WILEY-VCH Verlag GmbH & Co. KGaA, Weinheim. **c** Simulated architecture of nanocrystalline Si-interlayered perovskite-Si TSC, **d** cross-sectional FESEM image of the simulated stack, and **e** IPCE spectrum obtained from top and bottom cells, Reproduced from Ref. Mazzarella et al. (2019), Copyright (2019) with permission of WILEY-VCH Verlag GmbH & Co. KGaA, Weinheim

primary function is to enable the balanced photogenerated carriers' recombination, ensuring the continuous current flow across the whole tandem device. The 2T tandem structure's total photocurrent must match the photocurrents from both sub-cells to function. The sub-cell with the lower current is ultimately responsible for limiting the photocurrent. The entire tandem device's performance is not hindered by forming a p–n junction since the ICL functions as a recombination site, promoting efficient current flow.

Additionally, the ICL quality significantly affects the tandem device's V_{OC}. Incorporating an ICL with poor performance can result in charge carriers accumulating up at the ICL's interfaces. This accumulation introduces a reverse electric field, reducing the device's total output voltage. Therefore, it is crucial to optimize the ICL's various properties, including transmittance, thickness, resistivity, and refractive index, which is of paramount importance. The transmittance of the ICL affects the amount of incident light reaching the sub-cells, while the thickness and resistivity impact the electrical properties and charge carrier transport within the tandem device. Moreover, the refractive index of the ICL influences the optical matching and impedance matching between the sub-cells, affecting the overall device performance. By carefully optimizing these properties, researchers can enhance the performance of the ICL, thereby improving the efficiency and voltage output of the 2T TSC configuration.

In a recent study by Mazzarella et al., an interlayer comprising nanocrystalline silicon oxide was introduced between the perovskite and Si sub-cells to mitigate losses associated with infrared reflection. The interlayer's thickness and the refractive index were optimized by adjusting the oxygen content. The fabrication of 2T PSTSCs with a certified PCE of 25.2% was made possible by this optimization process. However, despite the promising PCE, the Si sub-cell at the bottom of the tandem device hampered performance. The bottom Si sub-cell exhibited a slightly lower J_{SC} of 18.8 mA cm^{-2} than the top perovskite sub-cell with a J_{SC} of 19.9 mA cm^{-2}. These current density (J_{SC}) values were determined from the external quantum efficiency (EQE) spectrum analysis (Fig. 5.6c–e). It is reasonable to expect improvements by attaining a better match in the J_{SC} values of both the bottom and top sub-cells to further enhance the photovoltaic performance of this type of tandem device. Addressing this mismatch in J_{SC} values could involve optimizing the device structure, materials, or interface engineering techniques. Researchers can aim for improved performance and higher efficiencies in this type of perovskite/silicon tandem devices by ensuring a more balanced current distribution between the sub-cells (Mazzarella et al. 2019).

According to Ashouri et al., an HTL that combines a *p–i–n* perovskite sub-cell with a SAM with a methyl group substitution [4-(3, 6-dimethyl-9H-carbazol-9-yl)butyl]phosphonic acid (Me-4PACz) can significantly improve hole extraction and transport efficiency. Using this strategy, a 2T PSTSC now holds the present certified PCE world record of 29.15%. Furthermore, there has been much interest in developing the charge-selective layers for the *n–i–p* architecture and the *p–i–n* configuration for the perovskite sub-cell. Researchers aim to enhance the performance of the charge-selective layers to enable efficient charge extraction and transport in the *n-i-p* perovskite sub-cell. Another notable strategy to improve stability involves incorporating a two-dimensional (2D) perovskite in conjunction with a three-dimensional

(3D) perovskite (Al-Ashouri et al. 2020). To create a 2D/3D mixed perovskite system, Duong et al. coated n-butylammonium bromide onto the surface of a 3D perovskite. This hybrid approach enhanced the charge-carrier lifetime, photovoltaic performance (with a PCE of 27.7%), and overall stability of the perovskite solar cell. These advancements in HTL design, charge-selective layers, and hybrid perovskite systems demonstrate the ongoing efforts to optimize and enhance the performance of perovskite/silicon tandem solar cells. The world record PCE achieved and the exploration of novel approaches signify the potential for achieving even higher efficiencies and improved stability in future generations of PSTSCs (Duong et al. 2020).

In another research investigation, the thin layer of cesium bromide (CsBr), which was thermally evaporated, was employed as an intermediate material between the perovskite layer and the HTL. This innovative approach led to fully-textured PSTSCs exhibiting remarkable performance, having a PCE of 27.48%. An unencapsulated configuration was utilized to assess the CsBr-based device's stability. The device was stored in darkness within a glove box filled with N_2 gas. Remarkably, the CsBr-based device demonstrated excellent stability for an extended period of over 10,000 h. The device maintained approximately 95% of its initial PCE throughout this duration. The control device without the CsBr layer exhibited significantly lower stability, with only 74% of its initial PCE remaining over the same timeframe. These findings highlight the effectiveness of employing a CsBr thin layer in enhancing the stability of PSTSCs. The results indicate that the CsBr-based device exhibited superior long-term performance, preserving a higher proportion of its original PCE than the control device. This suggests that introducing CsBr as an intermediate layer helps perovskite solar cells maintain their stability and lifetime (Li et al. 2021b).

Even though the comprehensive device structure and detailed experimental procedures hadn't been explicitly provided, the study's outcomes imply the potential to achieve the significant milestone of 30% PCE in PSCs. This achievement could be a crucial catalyst for driving the potential commercialization of tandem devices. By successfully attaining a PCE of 27.48% through the utilization of a thermally evaporated CsBr thin-layer, the study provides compelling evidence that further optimization and refinement of device architecture may enable researchers to breach the 30% PCE threshold. Such an advancement would have profound implications for the practical implementation and scalability of tandem devices, facilitating their integration into commercial applications. While additional research is necessary to fully elucidate the specific device configuration and optimize the fabrication process, the progress underscores the immense potential of perovskite solar cells and hints at the feasibility of achieving the ambitious 30% PCE target. These findings encourage further exploration and innovation in PSCs, fostering the pursuit of highly effective and commercially viable tandem devices.

5.3 Building-Integrated Photovoltaics (BIPVs)

Solar panels for electricity generation are commonly installed on building rooftops. However, the integration of photovoltaic systems into the construction of buildings, known as BIPV, is gaining attraction and being employed in innovative ways. BIPV incorporates PV panels into the building envelope, like skylights, windows, facades, or exterior walls. In contrast to rooftop-mounted systems, BIPV systems have different specifications emphasizing optical characteristics like color and transparency and weight and form factor considerations. PSCs offer unique advantages for BIPV applications. They are highly appealing because they can be synthesized on conductive, transparent, and flexible substrates. Three main kinds of perovskite-based BIPV have been produced and researched. The first group consists of semi-transparent PSCs that have been investigated for usage in glass roofs and building windows. These cells allow partial transmission of light while generating electricity.

The second category comprises colorful PSCs, which find application in building fences, walls, and car park roofs. These cells offer the ability to customize colors and integrate aesthetically pleasing designs into the building's exterior. Smart PV windows (SPWs) combine solar harvesting with electrochromic/thermochromic functions and fall under the third category of dual-functional BIPVs. On hot summer days, SPWs not only serve as a means to block excessive sunlight from entering the building but also have the intriguing capability of harvesting sunlight to generate electricity (Fig. 5.7a) (Bati et al. 2023). Nevertheless, before BIPV can be more widely used, several challenges related to manufacturing and performance need to be addressed. In the following subsections, we will detail and examine some significant recent achievements in these three areas of BIPV using PSCs. These advancements demonstrate the potential of PSCs in BIPV applications and contribute to the ongoing efforts to overcome existing challenges. By leveraging the unique properties of perovskite materials, researchers aim to enhance the performance, reliability, and manufacturability of BIPV systems, paving the way for their broader adoption in the construction industry (Shi et al. 2020).

5.3.1 Smart Photovoltaic Windows (SPWs)

A smart window refers to a type of glass that undergoes a change in its optical transmission characteristics when subjected to external stimuli such as heat, voltage, or light. Switchable films, such as thermochromic, photochromic, or electrochromic layers, are used to construct smart windows. Integrating smart windows with PV devices holds considerable potential for enhancing the energy efficiency of buildings. Several benefits can be achieved by combining the functionality of smart windows with PV technology. These include reducing cooling and heating costs, minimizing ventilation loads, improving privacy, and the ability to harvest surplus solar energy as electricity. Integrating smart windows and PV devices enables smart windows

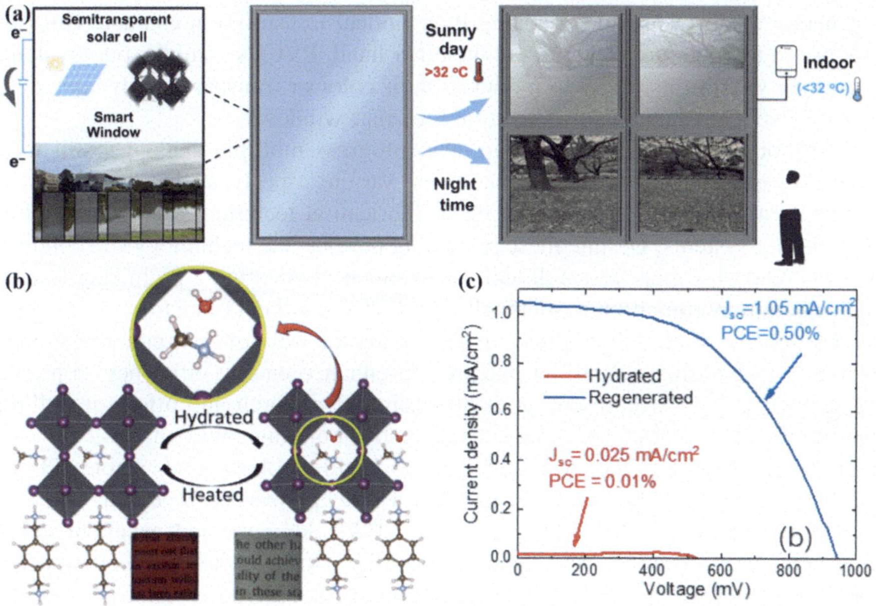

Fig. 5.7 **a** Schematic working of SPWs on sunny days and nights, Reproduced from Ref. Bati et al. (2023), Copyright (2023) with permission of Springer Nature. **b**, **c** Hydrated and regenerated schematic of a 2D perovskite material, namely $(C_6H_4(CH_2NH_3)_2)(CH_3NH_3)[Pb_2I_7]$ and its performance as SPW, Reproduced from Ref. Zhang et al. (2021b), Copyright (2021) with permission of the American Chemical Society

to actively modulate their optical properties while generating electrical power from excess sunlight. This synergistic approach maximizes the overall energy efficiency of the building. By dynamically controlling the optical transmission of the smart windows, the need for additional cooling or heating can be reduced, resulting in energy savings. Additionally, the integration of PV devices allows for the conversion of excess solar energy into electricity, further enhancing the sustainability and efficiency of the building (Wang et al. 2023).

The crystallization process of perovskite light absorbers, which requires specific temperatures, presents intriguing possibilities for research in SPWs. The temperature manipulation can be utilized to adjust the color of perovskite films, opening new avenues for exploration and innovation. The development of integrated SPW systems that combine a PV device with an electrochromic layer has seen significant advancements. Various types of developing PV cells, including organic solar cells, DSSCs, and PSCs, have been investigated for their integration into SPWs. Among these, PSCs hold particular appeal due to their high voltage output, which is crucial for operating electrochromic windows effectively. Based on their functionality, the two main types of perovskite-based SPWs are classified into photovoltachromic cells (PVCCs) and dual-function thermochromic solar cells. Dual-function thermochromic solar cells exhibit solar-energy harvesting and thermochromic properties, enabling them to generate electricity while adjusting their optical transmission characteristics in response to temperature changes. On the other hand, PVCCs combine photovoltaic functionality with the ability to modulate their color or transparency dynamically, offering a versatile and adaptive solution for smart windows.

Even though this field has made significant progress, multiple challenges still need to be addressed. These challenges encompass various aspects, such as performance optimization, long-term stability, scalable fabrication techniques, and integration into building systems. Despite these remaining hurdles, the technology surrounding SPWs is ready for more research and advancement, indicating a promising future. Dual-function thermochromic solar cells and PVCCs will continue to be researched and developed, which will progress SPWs, paving the way for their practical implementation in buildings. With the potential to enhance energy efficiency, comfort, and aesthetics, SPWs represent an exciting area of research and offer compelling opportunities for creating sustainable and adaptive building environments.

In recent studies, the development of highly robust and stable SPWs has been achieved through the use of a 2D perovskite material, namely $(C_6H_4(CH_2NH_3)_2)(CH_3NH_3)[Pb_2I_7]$. These SPWs demonstrated excellent durability, remarkable stability, and rapid reversibility. However, their PV efficiencies were less than 1% (Fig. 5.7b, c) (Zhang et al. 2021b). The thermochromic perovskite-based smart solar windows field has faced several vital challenges recently. One of the main challenges is the relatively low efficiency of the devices, limiting their ability to generate significant amounts of electrical energy from sunlight. Additionally, the poor reversibility of the thermochromic transition, meaning the window's ability to return to its original state after being heated or cooled, is a challenge that needs to be addressed. Furthermore, ensuring the long-term stability of these SPWs remains a critical concern to ensure their practical application.

In addition to these challenges, an intriguing research direction is the exploration of lead-free perovskite materials for thermochromic SPWs. Investigating the thermochromic characteristics of lead-free perovskites holds promise for addressing environmental concerns and expanding the range of materials available for SPW applications. The vital objectives in advancing thermochromic perovskite-based smart solar windows must prioritize addressing poor device efficiency, improving reversibility, and enhancing long-term stability. Concurrently, investigating lead-free perovskites for their thermochromic properties presents an exciting avenue for further research. By addressing these challenges and exploring alternative materials, SPWs can progress toward more efficient, reversible, and environmentally friendly solutions with improved performance and broader practical applications.

5.4 Perovskite Solar Cells for Space Applications

PSCs exhibit distinct properties that make them highly promising for space applications. Notably, compared to other solar cell technologies, they possess a high power-to-weight ratio (specific power) and exceptional resistance to gamma radiation (Wheeler et al. 2017). Moreover, the issues of instability encountered by PSCs on Earth due to exposure to oxygen and moisture are eliminated in the space environment, besides enhancing the potential of PSCs for space-based use. It is appropriate to examine the effectiveness of solar cells for space applications using the AIAA-S111 standard, which outlines the qualifying requirements for space solar cells (Wheeler et al. 2017). The solar cell system should fulfill specific criteria related to performance and stability before considering its suitability for space missions. For example, solar cells must withstand exposure to 1 MeV electrons at a fluence of 1×10^{16} electrons per square centimeter and 3 MeV protons at a fluence of 1×10^{13} protons per square centimeter.

Additionally, a temperature range of between − 150 and 150 °C should be encompassed in the characterization of solar cells. Recently, the maximum achieved PCE for PSCs is 47.1%, accomplished when six-junction inverted metamorphic solar cells with 143 Suns of light are employed (Fig. 5.8a, b) (Geisz et al. 2020). By the IEC-61345 industrial standard, the preservation of the initial PCE in a solar cell system is crucial. Specifically, the standard requires that the PCE remains above 95% after exposure to 15 kWh m^{-2} of UV radiation. In space applications, the significance of UV exposure becomes even more pronounced, given that the intensity of UV radiation in space exceeds that experienced under standard AM0 (Air Mass Zero) conditions. Consequently, subjecting PSCs to rigorous testing under high UV irradiance is crucial. Simulated space environments are commonly employed in the performance testing of PSCs due to the challenging and intricate nature of conducting such tests under actual space conditions. It is crucial to assess the survival capacity of PSCs in various space conditions, including exposure to high-energy particle illumination (such as protons, electrons, and gamma rays), high vacuum conditions, and high temperatures. Encouraging test outcomes have been observed for single-junction PSCs, indicating their potential suitability for space applications (Chen et al. 2018).

In a recent study by Lang et al., the durability of all-perovskite-tandem devices was investigated under conditions of high-energy proton illumination. The devices were irradiated at a proton energy of 68 MeV, with an accumulative dosage of 1×10^{13} p^+/cm^2. Notably, the study revealed that over 94% of the original PCE was maintained, validating the robustness of perovskite materials against high irradiation exposure (Fig. 5.8c, d). This finding establishes perovskite materials as viable candidates for applications in the space industry. Notably, the accumulated dose of 1×10^{13} p^+/cm^2 used in the study is identical to the accumulated dose expected over a period of > 100 years in near-earth orbit and > 10 years in geostationary orbit. Even though PSC development for space applications is in the early stages, a few studies have been conducted under actual space conditions (Lang et al. 2020). The pioneering work undertaken by Cardinaletti et al. stands out, as they were the first to monitor

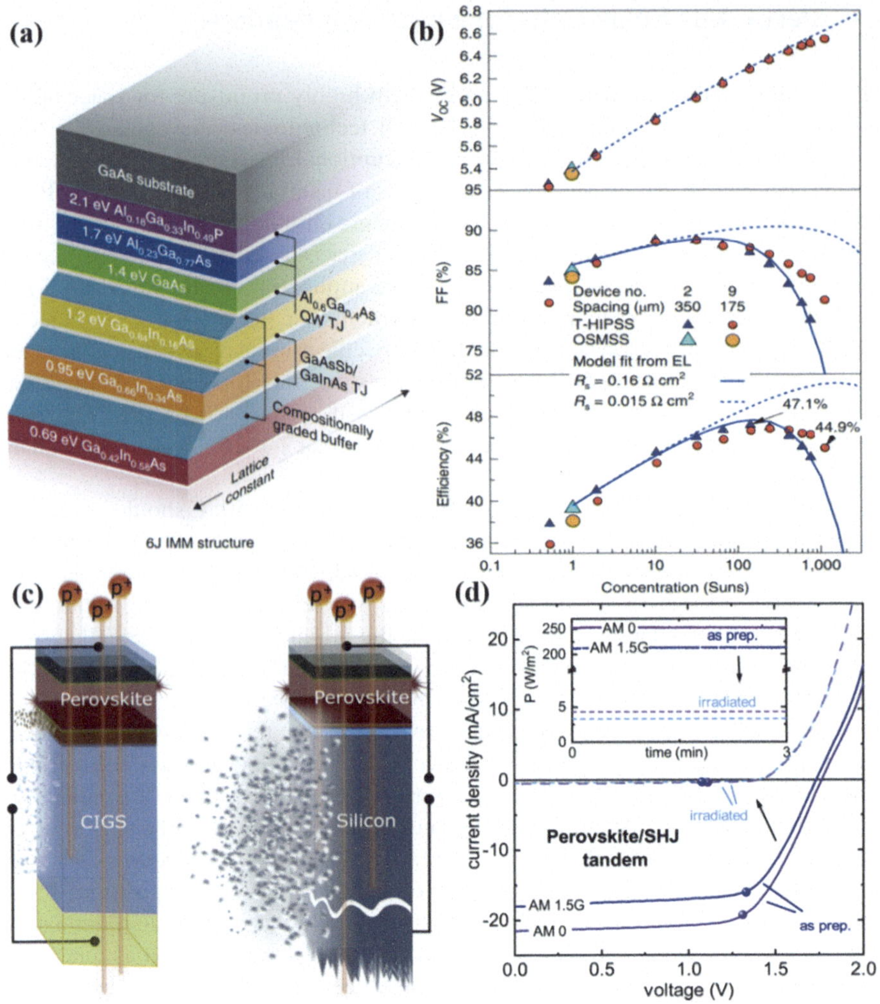

Fig. 5.8 **a** Schematics of six-junction inverted metamorphic TSC under the illumination of 143 Suns, **b** photovoltaic performances such as V_{OC}, FF, and PCE, Reproduced from Ref. Geisz et al. (2020), Copyright (2020) with permission of Springer Nature. **c** Schematics demonstrating the robustness of perovskite materials against high irradiation exposure, and **d** stability testing confirms that 94% of initial PCE is retained, Reproduced from Ref. Lang et al. (2020), Copyright (2020) with permission of Elsevier

the performance changes of PSCs deployed on stratospheric balloons that rose to an altitude of 32 km. Throughout the approximately 3-h stratospheric flight, the performance of PSCs based on $MAPbI_3$ was measured in the near-space environment. While this study provided valuable insights, it is essential to conduct longer-duration tests to better understand PSCs' behavior in actual space environments (Cardinaletti et al. 2018).

In another study, Tu et al. employed a high-altitude balloon to transport a large-area PSC (1.00 cm^2) based on $FA_{0.81}MA_{0.10}Cs_{0.04}PbI_{2.55}Br_{0.40}$ to an altitude of 35 km for a duration of 2 h. Remarkably, throughout the test, the TiO_2-based PSC exhibited a preservation of 95% of its initial PCE. The authors also found that incorporating a UV filter might enhance the stability of these devices even further. These results have established a strong basis for further investigation to advance the utilization of PSCs in space applications. It should be noted, however, that while significant progress has been made in investigating the viability of perovskite-based devices in actual space conditions, these efforts have primarily focused on single-junction PSCs. Comparable endeavors to explore the practical application of perovskite-based tandem solar cells in space still need to be reported (Tu et al. 2019).

5.5 PV-Based Energy Storage Systems

Solar energy is a leading renewable energy source, but traditional solar cells have limitations as they generate electricity instantaneously, requiring immediate consumption or storage. There is a growing need for integrated energy conversion-storage systems to ensure continuous energy availability. A promising solution involves integrating solar cells with energy storage units, enabling renewable and clean technologies to be developed. Among the various energy storage options, supercapacitors (SCs), lithium-ion batteries (LIBs), and various rechargeable batteries are highly intriguing because of their high energy and power density and longer lifetimes. Integrated systems combining these energy storage devices with different solar cell technologies like DSSCs and OPVs have received considerable attention (Zeng et al. 2020). However, these integrated systems' overall performance is still subpar, primarily because of the solar cells' low PCEs. Further advancements are needed to enhance the efficiency and effectiveness of these integrated energy conversion-storage systems.

The cost–benefit analysis is crucial to investment decisions in PV-integrated energy storage systems. Evaluating cost-effectiveness requires accounting for various factors, including system architecture, component sizes (energy storage devices, PV modules, electric cables, inverters, etc.), operation and maintenance costs, and potential substitutions. It is essential to ensure that integrated systems' cost-effectiveness exceeds the costs associated with the technology to realize their benefits. Manufacturing costs and PCE of individual PV cells are critical for evaluating the overall cost advantages of PV-integrated energy systems. Figure 5.9a presents a schematic representation showcasing PV-integrated energy storage devices and PV-cell-driven catalytic reactions to emphasize the benefits of integrated systems. The figure also provides information on the average voltage outputs necessary for driving various applications, including supercapacitors, water splitting, CO_2 reduction, and batteries. In the following sections, we will comprehensively describe and assess the latest advancements in integrated systems that combine PSCs with energy storage devices like batteries and supercapacitors. These evaluations aim to shed light on the developments achieved in this field and highlight the potential of such integrated systems for various energy-related applications (Bati et al. 2023).

As eco-friendly portable power supply technologies, photosupercapacitors, which consist of a PSC integrated with a supercapacitor, have gained significant attention. A distinctive feature of these devices' integration is their ability to operate without external wire connections. However, a considerable challenge in their implementation is high operating and output voltages. To address this challenge, Liu et al. have designed a system by assembling four separate photosupercapacitors in series, achieving a stable output voltage of approximately 3.8 V. After being charged through photovoltaic means, this innovative power pack successfully powered light-emitting diodes (LEDs), showcasing the capability of this technology. Supercapacitors require

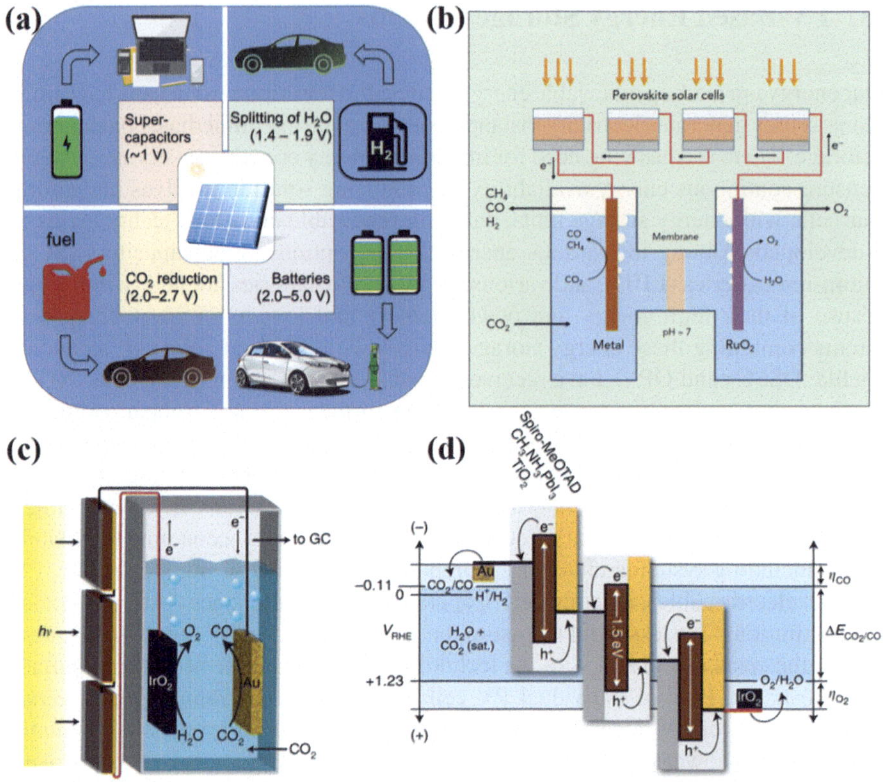

Fig. 5.9 **a** Schematic representation of PV-driven energy storage systems, catalysis reactions, etc., Reproduced from Ref. Bati et al. (2023), Copyright (2023) with permission of Springer Nature. **b** Typical schematic illustration of the CO_2 reduction reaction, Reproduced from Ref. Esiner et al. (2020), Copyright (2020) with permission of Elsevier. **c**, **d** Schematic working and energy diagram illustrating the process of converting CO_2 into CO using a series of three PSCs. Reproduced from Ref. Schreier et al. (2015), Copyright (2015) with permission of Springer Nature

high voltage; therefore, combining them with PSCs to create high-efficiency photo-supercapacitors holds promise as solar rechargeable supercapacitors (Liu et al. 2017). In addition, the proliferation of smart electronic devices, electric vehicles, and smart grids has garnered significant attention, indicating their potential integration into our daily lives. Rechargeable batteries are crucial to these cutting-edge technologies' energy storage systems. Among the various battery options available, metal-ion batteries, such as LIBs, stand out because of their excellent chemical stability and high energy density. In terms of energy storage for solar rechargeable batteries, LIBs are considered promising. Early reports have shown how LIBs can be integrated with DSSCs and silicon-based multi-junction solar cells. These developments showcase the potential synergy between LIBs and different types of solar cells, paving the way for efficient solar rechargeable battery systems.

As previously discussed, numerous perovskite materials have been developed, presenting potential opportunities for their integration with batteries. However, it is imperative to conduct comprehensive lifetime tests that encompass light, thermal, moisture, and stability evaluations under challenging testing environments for prolonged periods to commercialize these integrated systems successfully. These tests will provide crucial insights into the long-term performance and durability of the integrated systems. Despite significant advancements in integrated PSC-battery systems, there is still a need to minimize wire connections in future research endeavors. This step is essential to mitigate energy losses and reduce device fabrication costs. By minimizing wire connections, the overall efficiency of the integrated systems can be enhanced, thereby contributing to improved system performance and economic viability. Future work should prioritize the development of innovative strategies to achieve seamless integration between PSCs and batteries, aiming for efficient and cost-effective designs.

5.6 Perovskite Solar Cell-Based Catalysis

Solar-powered catalytic reactions are considered a burgeoning and environmentally friendly method for sustainable chemical production. Among the various types of solar-driven catalysis, such as photocatalytic, photoelectrochemical catalytic, and photothermal catalytic, as well as photosynthetic processes, advanced catalytic reactions like water splitting and carbon dioxide (CO_2) reduction offer promising avenues for generating renewable energy and providing cost-effective feedstocks for industries. These reactions utilize solar energy to drive chemical transformations, enabling greener and more sustainable energy production and raw material synthesis routes (Huang et al. 2020).

Hydrogen (H_2) energy, or H_2 fuel, is a well-known zero-carbon fuel that finds applications in fuel cells and internal combustion engines. It can be easily distributed through already-existing natural gas pipelines and conveniently stored as a compressed gas. One promising approach for producing H_2 in an environmentally friendly manner is through water (H_2O) electrolysis. This process involves the application of an electric current to a suitable electrode, resulting in the splitting of H_2O into H_2 and oxygen (O_2). Efficient electrocatalysts and high-voltage photovoltaic-electrolyzer systems play a crucial role in water electrolysis. Theoretically, a minimum energy input of 1.23 V, known as the thermodynamic equilibrium potential, is required for water splitting. However, practical operating potentials typically range between 1.5 and 2.0 V (Li et al. 2021a). Multiple junctions and individual cells must be coupled in series within the electrolyzer to achieve these high operating voltages. The electrochemical potential necessary for water splitting can be efficiently met by using two individual PSCs with V_{OC} values greater than 1.0 V.

A practical and economically viable approach for developing a water-splitting system powered by PV cells involves utilizing high-voltage solar cells. Among the potential candidates, bromine (Br)-based cells and single-junction all-inorganic PSCs have shown promise. Another appealing option is the construction of perovskite-based tandem PV devices, particularly the 2T configuration, which can be integrated into water-splitting systems. Notably, Bin et al. reported the first hybrid PTSC-driven water-splitting system, employing graphene-based catalysts for both H_2 and O_2 evolution. Their tandem cell achieved a V_{OC} value of 1.86 V, fulfilling the water-splitting voltage requirement. However, the achieved solar-to-hydrogen (STH) efficiency was below 10% (Bin et al. 2016). Several other reports are available related to the PSC-driven water splitting, and continuous progress is going on. Despite achieving a significant milestone, there is a need for additional enhancements to ensure their economic viability in both the stability and STH performance of the integrated water-splitting systems. Presently, the consumed power (electricity) holds a more significant value than the H_2 produced. Further advancements in integrated systems are envisaged due to the optimization of tandem solar cells' PV performance and the development of effective bifunctional electrocatalysts. These improvements aim to enhance the overall efficiency of the system and enable a more cost-effective utilization of solar energy for water splitting.

Other than PSC-driven water splitting, using catalysis research to convert CO_2 into beneficial chemical feedstocks and fuels remains a long-standing focus. This approach holds significant importance for mitigating the global greenhouse effect caused by CO_2 emissions and producing high-value chemicals. The CO_2 reduction reaction (CO_2RR), as shown in Fig. 5.9b, driven by solar-generated renewable electricity, offers a promising pathway to utilize these abundant resources to synthesize valuable chemicals efficiently. However, the CO_2RR process requires driving voltages surpassing single-junction solar cells' capabilities. Precisely, the power supply unit must specifically deliver an output voltage exceeding 2.0 V, necessitating the series connection of single-junction PV cells to achieve the desired voltage levels (Esiner et al. 2020). Figure 5.9c, d depicts the energy diagram illustrating the process of converting CO_2 into CO using a series of three PSCs. A more recent study developed a system for light-driven CO_2 conversion to CH_4, wherein four series-connected PSCs were employed. These PSCs exhibited a V_{OC} of 4.20 V. An electrochemical cell containing Cu and RuO_2 electrodes was electrically connected to the four PSCs, resulting in a 5% efficiency for converting solar energy into fuel (Schreier et al. 2015).

Given the requirement for high driving voltages in the CO_2RR, it is rational to anticipate that attaining high-efficiency CO_2 conversion at a low cost would involve utilizing multi-junction solar cells, specifically triple-junction PSCs. It is important to note that despite numerous research groups reporting high-efficiency perovskite triple-junction solar cells, to the best of our knowledge, there has been no previous endeavor to design perovskite multi-junction (tandem) solar cells explicitly for CO_2RR.

5.7 Conclusion

Metal halide perovskites have become highly intriguing PV materials because of their remarkable properties, such as high absorption coefficients, tunable bandgaps, superior charge-carrier mobilities, and solution processability. These materials have witnessed unprecedented advancements in PCE in PV devices, representing the most rapid growth among all PV technologies in history. PSCs are now perfect candidates for system integration due to the fantastic advancements in PV performance, which will allow for the realization of cutting-edge technologies in the next generation.

The next-generation perovskite-based solar cells have different applications that can encompass various areas. Tandem PV cells, for instance, involve combining perovskite solar cells with other PV technologies to achieve higher efficiency and broader spectral coverage. Space applications leverage the exceptional properties of perovskite-based solar cells to provide efficient and lightweight power generation solutions for space missions. Perovskite solar cells are integrated with energy storage components in PV-integrated energy storage systems, enabling simultaneous energy generation and storage. PV cell-driven catalysis utilizes perovskite solar cells as efficient light sources for driving catalytic reactions, facilitating sustainable chemical transformations. Building-integrated photovoltaics (BIPVs) involve incorporating perovskite solar cells into building materials, enabling seamless integration of energy generation within architectural structures. Perovskite-based solar cells are highly promising for next-generation applications because of the exciting properties of metal halide perovskites and their significant progress in PV performance. Their potential integration with other systems paves the way for advancements in BIPVs, tandem PV cells, PV-integrated energy storage systems, PV cell-driven catalysis, and space applications. These emerging applications demonstrate the transformative potential of perovskite-based solar cells in driving innovative technologies and shaping the future of renewable energy.

Tandem solar cells made of PSCs and other PV cells have shown significant potential for realizing high-efficiency PV systems and leveraging existing PV technologies. However, there are significant challenges that require immediate attention. Low-bandgap perovskites should be used when integrated using wide-bandgap semiconductor-based cells. However, the development of low-bandgap perovskites, often accomplished by partially substituting Pb^{2+} with Sn^{2+}, introduces high defect densities, pinholes, uneven surfaces, and decreased carrier lifetime and diffusion length. Alternative strategies, which include substituting other stable metals for Sn and applying surface passivation methods, should be explored to design low-band perovskites. On the other hand, when pairing with low-bandgap materials like CdTe as the top sub-cell, wide-bandgap perovskites are necessary. Wide-bandgap perovskites currently have lower efficiencies, which must be fixed to reach high PV efficiencies. Moreover, highly conductive transparent electrodes, such as carbon-based materials, must be further developed for semi-transparent top subcells in tandem solar cells. The top and bottom sub-cells photocurrent matching is crucial for achieving high PV performance, and to increase their chances of commercialization, tandem solar cells' operational stability should be tested in harsher conditions for a longer duration of time.

PSCs have high output voltage levels, which offers new opportunities for integrated energy storage systems. Integrating PSCs with batteries or supercapacitors in solar rechargeable power systems necessitates voltage matching between the energy conversion unit and the storage component. The voltage requirements for energy storage systems can be met by perovskite-based multi-junction tandem solar cells with greater voltages acquired by combining multiple series-connected cells. Energy storage device construction can benefit significantly from using carbon-based materials because of their inexpensiveness and strong conductivity. Integrated systems that utilize carbon electrode-based PSCs and energy storage devices hold major significance. Additionally, exploring various perovskite structures beyond the conventional $MAPbI_3$ can further enhance the development of PSC-integrated energy storage systems. The optimization of electrocatalysts for PV-driven conversion of CO_2 and water splitting is critical, and tandem solar cells can provide the required high driving voltages. However, the power consumed in PV cells should be considered for CO_2 reduction reactions. Progress in semi-transparent perovskite solar cells for BIPVs is encouraging, but improvements are needed in both PCE and optical transparency. Strategies such as compositional and bandgap engineering should be explored to achieve higher photovoltage values and compensate for lower photocurrent in semi-transparent and colorful PSCs, aiming for the benchmark AVT value of 25% for solar windows.

In summary, integrated PSCs with various PV cells offer opportunities for cutting-edge technologies and highly effective PV systems. The advancement and commercialization of perovskite-based solar cells in multiple applications will be aided by addressing challenges related to materials design, electrode development, photocurrent matching, operational stability, voltage matching in energy storage systems, catalytic activity, and improving BIPV performance, will contribute to the advancement and commercialization of perovskite-based solar cells in various applications.

References

Al-Ashouri A, Magomedov A, Roß M et al (2019) Conformal monolayer contacts with lossless interfaces for perovskite single junction and monolithic tandem solar cells. Energy Environ Sci 12:3356–3369. https://doi.org/10.1039/c9ee02268f

Al-Ashouri A, Köhnen E, Li B et al (2020) Monolithic perovskite/silicon tandem solar cell with >29% efficiency by enhanced hole extraction. Science (80-) 370:1300–1309. https://doi.org/10.1126/science.abd4016

Barkhouse DAR, Gunawan O, Gokmen T et al (2015) Yield predictions for photovoltaic power plants:empirical validation, recent advances and remaining uncertainties. Prog Photovoltaics Res Appl 20:6–11. https://doi.org/10.1002/pip

Bati ASR, Zhong YL, Burn PL et al (2023) Next-generation applications for integrated perovskite solar cells. Commun Mater 4:1–24. https://doi.org/10.1038/s43246-022-00325-4

Bin AR, Yusoff M, Jang J (2016) Highly efficient photoelectrochemical water splitting by a hybrid tandem perovskite solar cell. Chem Commun 52:5824–5827. https://doi.org/10.1039/c6cc01249c

Brinkmann KO, Becker T, Zimmermann F et al (2022) Perovskite–organic tandem solar cells with indium oxide interconnect. Nature 604:280–286. https://doi.org/10.1038/s41586-022-04455-0

Cardinaletti I, Vangerven T, Nagels S et al (2018) Organic and perovskite solar cells for space applications. Sol Energy Mater Sol Cells 182:121–127. https://doi.org/10.1016/j.solmat.2018.03.024

Chen CC, Bae SH, Chang WH et al (2015) Perovskite/polymer monolithic hybrid tandem solar cells utilizing a low-temperature, full solution process. Mater Horizons 2:203–211. https://doi.org/10.1039/c4mh00237g

Chen W, Zhang J, Xu G et al (2018) A semitransparent inorganic perovskite film for overcoming ultraviolet light instability of organic solar cells and achieving 14.03% efficiency. Adv Mater 30:1–10. https://doi.org/10.1002/adma.201800855

Chen X, Jia Z, Chen Z et al (2020) Efficient and reproducible monolithic perovskite/organic tandem solar cells with low-loss interconnecting layers. Joule 4:1594–1606. https://doi.org/10.1016/j.joule.2020.06.006

De Vos A et al (1980) Detailed balance limit of the efficiency of tandem solar cells. J Phys D Appl Phys 13:839–846. https://doi.org/10.1088/0022-3727/13/5/018

Duong T, Pham H, Kho TC et al (2020) High efficiency perovskite-silicon tandem solar cells: effect of surface coating versus bulk incorporation of 2d perovskite. Adv Energy Mater 10:1–15. https://doi.org/10.1002/aenm.201903553

Esiner S, Wang J, Janssen RAJ (2020) Light-driven electrochemical carbon dioxide reduction to carbon monoxide and methane using perovskite photovoltaics. Cell Reports Phys Sci 1. https://doi.org/10.1016/j.xcrp.2020.100058

Geisz JF, France RM, Schulte KL et al (2020) Six-junction III–V solar cells with 47.1% conversion efficiency under 143 Suns concentration. Nat Energy 5:326–335. https://doi.org/10.1038/s41560-020-0598-5

Huang H, Pradhan B, Hofkens J et al (2020) Solar-driven metal halide perovskite photocatalysis: design, stability, and performance. ACS Energy Lett 5:1107–1123. https://doi.org/10.1021/acsenergylett.0c00058

Jošt M, Bertram T, Koushik D et al (2019) 21.6%-efficient monolithic perovskite/Cu(In, Ga)S_e2 tandem solar cells with thin conformal hole transport layers for integration on rough bottom cell surfaces. ACS Energy Lett 4:583–590. https://doi.org/10.1021/acsenergylett.9b00135

Jošt M, Köhnen E, Al-Ashouri A et al (2022) Perovskite/CIGS tandem solar cells: from certified 24.2% toward 30% and beyond. ACS Energy Lett 7:1298–1307. https://doi.org/10.1021/acsenergylett.2c00274

Lang F, Jošt M, Frohna K et al (2020) Proton radiation hardness of perovskite tandem photovoltaics. Joule 4:1054–1069. https://doi.org/10.1016/j.joule.2020.03.006

Li TT, Yang YB, Li GR et al (2021a) Two-terminal perovskite-based tandem solar cells for energy conversion and storage. Small 17:1–24. https://doi.org/10.1002/smll.202006145

Li Y, Shi B, Xu Q et al (2021b) Wide bandgap interface layer induced stabilized perovskite/silicon tandem solar cells with stability over ten thousand hours. Adv Energy Mater 11:1–9. https://doi.org/10.1002/aenm.202102046

Liu Y, Renna LA, Bag M et al (2016) High efficiency tandem thin-perovskite/polymer solar cells with a graded recombination layer. ACS Appl Mater Interfaces 8:7070–7076. https://doi.org/10.1021/acsami.5b12740

Liu Z, Zhong Y, Sun B et al (2017) Novel integration of perovskite solar cell and supercapacitor based on carbon electrode for hybridizing energy conversion and storage. ACS Appl Mater Interfaces 9:22361–22368. https://doi.org/10.1021/acsami.7b01471

Mazzarella L, Lin YH, Kirner S et al (2019) Infrared light management using a nanocrystalline silicon oxide interlayer in monolithic perovskite/silicon heterojunction tandem solar cells with efficiency above 25%. Adv Energy Mater 9:1–9. https://doi.org/10.1002/aenm.201803241

NREL (2023) Best research-cell efficiency chart

Palmstrom AF, Eperon GE, Leijtens T et al (2019) Enabling flexible all-perovskite tandem solar cells. Joule 3:2193–2204. https://doi.org/10.1016/j.joule.2019.05.009

Pan Y, Zheng X, Guo J, Chen Z, Li S, He C, Ye S, Xia X, Wang S, Lu X, Zhu H, Min J, Zuo L, Shi M, Chen H (2022) A new end group on nonfullerene acceptors endows efficient organic solar cells with low energy losses. Adv Funct Mater 32:2108614. https://doi.org/10.1002/adfm.202108614

Schreier M, Curvat L, Giordano F et al (2015) Efficient photosynthesis of carbon monoxide from CO_2 using perovskite photovoltaics. Nat Commun 6:1–6. https://doi.org/10.1038/ncomms8326

Shen H, Walter D, Wu Y et al (2020) Monolithic perovskite/Si tandem solar cells: pathways to over 30% efficiency. Adv Energy Mater 10:1–18. https://doi.org/10.1002/aenm.201902840

Shi B, Duan L, Zhao Y et al (2020) Semitransparent perovskite solar cells: from materials and devices to applications. Adv Mater 32:1–12. https://doi.org/10.1002/adma.201806474

Tu YG, Xu GN, Yang XY et al (2019) Mixed-cation perovskite solar cells in space. Sci China Physics, Mech Astron 62. https://doi.org/10.1007/s11433-019-9356-1

Wang Z, Zhu X, Zuo S et al (2020) 27%-efficiency four-terminal perovskite/silicon tandem solar cells by sandwiched gold nanomesh. Adv Funct Mater 30:1–8. https://doi.org/10.1002/adfm.201908298

Wang P, Li W, Sandberg OJ et al (2021) Tuning of the interconnecting layer for monolithic perovskite/organic tandem solar cells with record efficiency exceeding 21%. Nano Lett 21:7845–7854. https://doi.org/10.1021/acs.nanolett.1c02897

Wang Y, Chen Q, Liu Z et al (2023) Acidochromic organic photovoltaic integrated device. Chem Eng J 452:139479. https://doi.org/10.1016/j.cej.2022.139479

Wheeler LM, Moore DT, Ihly R et al (2017) Switchable photovoltaic windows enabled by reversible photothermal complex dissociation from methylammonium lead iodide. Nat Commun 8. https://doi.org/10.1038/s41467-017-01842-4

Yakushev MV, Mudryi AV, Gremenok VF et al (2004) Influence of growth conditions on the structural quality of $Cu(InGa)Se_2$ and $CuInSe_2$ thin films. Thin Solid Films 451–452:133–136. https://doi.org/10.1016/j.tsf.2003.11.003

Yang Z, Yu Z, Wei H et al (2019) Enhancing electron diffusion length in narrow-bandgap perovskites for efficient monolithic perovskite tandem solar cells. Nat Commun 10:1–9. https://doi.org/10.1038/s41467-019-12513-x

Zeng Q, Lai Y, Jiang L et al (2020) Integrated photorechargeable energy storage system: next-generation power source driving the future. Adv Energy Mater 10:1–30. https://doi.org/10.1002/aenm.201903930

Zhang Y, Kirs A, Ambroz F et al (2021a) Ambient fabrication of organic-inorganic hybrid perovskite solar cells. Small Methods 5:1–40. https://doi.org/10.1002/smtd.202000744

Zhang Y, Wang Z, Hu S et al (2021b) Robust and swiftly reversible thermochromic behavior of a 2D perovskite of $(C_6H_4(CH_2NH_3)_2)(CH_3NH_3)[Pb_2I_7]$ for smart window and photovoltaic smart window applications. ACS Appl Mater Interfaces 13:12042–12048. https://doi.org/10.1021/acsami.1c00163

Zheng J, Wang G, Duan W et al (2022) Monolithic perovskite-perovskite-silicon triple-junction tandem solar cell with an efficiency of over 20%. ACS Energy Lett 7:3003–3005. https://doi.org/10.1021/acsenergylett.2c01556

Zhu Z, Mao K, Xu J (2021) Perovskite tandem solar cells with improved efficiency and stability. J Energy Chem 58:219–232. https://doi.org/10.1016/j.jechem.2020.09.022

The manufacturer's authorised representative in the EU is Springer Nature Customer Service Centre GmbH, Europaplatz 3, 69115 Heidelberg, Germany. If you have any concerns regarding our products, please contact ProductSafety@springernature.com

Printed and bound by CPI Group (UK) Ltd, Croydon, CR0 4YY
01/10/2026
02225923-0005